Intentional Oil Pollution at Sea

Global Environmental Accords Series
Nazli Choucri, editor

Global Accord: Environmental Challenges and International Responses
Nazli Choucri, editor

Institutions for the Earth: Sources of Effective International Environmental Protection
Peter M. Haas, Robert O. Keohane, and Marc A. Levy, editors

Intentional Oil Pollution at Sea: Environmental Policy and Treaty Compliance
Ronald B. Mitchell

Intentional Oil Pollution at Sea
Environmental Policy and
Treaty Compliance

Ronald B. Mitchell

The MIT Press
Cambridge, Massachusetts
London, England

This book was set in Sabon by DEKR Corporation and was printed and bound in the United States of America.

Library of Congress Cataloging-in-Publication Data

Mitchell, Ronald B. (Ronald Bruce)
  Intentional oil pollution at sea : environmental policy and treaty compliance / Ronald B. Mitchell.
    p. cm. — (Global environmental accords series)
  Includes bibliographical references and index.
  ISBN 0-262-13303-2
  1. Oil pollution of the sea—Law and legislation. 2. Oil pollution of the sea—Prevention. 3. Oil pollution of the sea—Government policy. 4. Treaties —Psychological aspects. I. Title. II. Series.
  K3590.4.M58 1994
  341.7′623—dc20   94–13537
                                                                         CIP

For Amy

Contents

Acknowledgments

This book has benefited immensely from the intellectual guidance and support of many people. Robert Keohane, Abram Chayes, William Clark, and Antonia Chayes have consistently given generously of their time, thought, and energy in reading, discussing, and critiquing numerous versions of the chapters. Feedback, comments, and suggestions from each of them have improved both the content and the form of the book. All have wholeheartedly joined me in my effort to understand the sources of environmental treaty compliance.

I have also received unparalleled assistance throughout the writing process from David Weil and Eileen Babbitt. Without their help, the book truly would not have been possible. Marc Levy, Edward Parson, and Vicki Norberg-Bohm have helped me enormously in understanding international environmental problems and the methodologies needed to distinguish between causes and effects. Dan Gilligan has provided valuable research assistance. Holly Tanner has provided excellent copyediting. Numerous other friends and colleagues have contributed to the research presented here by commenting on various drafts of chapters or through their involvement with me in common research projects. They include Harold Jacobson, Gail Osherenko, Edith Brown Weiss, Oran Young, and Mark Zacher. Ambassador Sidney Wallace and Mr. Bin Okamura, as well as numerous other officials at the International Maritime Organization (IMO) in London, generously allowed me to observe the thirty-first session of the Marine Environment Protection Committee, answered numerous questions, and gave me full use of the IMO library facilities. Officials of the U.S. Coast Guard, especially Commander William Chubb, generously answered my numerous questions and gave me access to various archival materials. I was aided in my research by comments made during presentations at Harvard University's International Environmental Institutions Research Seminar during 1991 and 1992. The final drafts of the manuscript have benefited greatly from the comments of three anonymous reviewers. Madeline Sunley, Ann Sochi, and Melissa Vaughn of the MIT Press helped me navigate the world of book publishing and copyediting.

Much of the background research for this book was done while I was a graduate student at Harvard University's Kennedy School of Government. I am grateful for research funding provided by Harvard's Center for International Affairs, Harvard's MacArthur Scholarships in International Security, and the Eisenhower World Affairs Institute. Summer travel funding was provided through the Center for International Affairs by the Delegation of the Commission of the European Communities and by the Rockefeller Brothers Fund. My research was conducted at the Center for International Affairs, Harvard University; the Center for Science and International Affairs, Harvard University; and the Department of Political Science, University of Oregon.

My parents, Lee Mitchell and Edith Bertsche Mitchell, and my five siblings have all contributed to this enterprise, each in a different way helping me develop and indulge an intellectual curiosity and a willingness to take the risks necessary to pursue and achieve my goals. Finally, and most important, Amy S. Reiss has consistently showered me with love, support, friendship, understanding, and a wonderful sense of priorities throughout the trials and tribulations that have accompanied the writing of this book.

I

Introduction

Too many people assume, generally without having given any serious thought to its character or its history, that international law is and always has been a sham. Others seem to think that it is a force with inherent strength of its own. . . . Whether the cynic or sciolist is the less helpful is hard to say, but both of them make the same mistake. They both assume that international law is a subject on which anyone can form his opinions intuitively, without taking the trouble, as one has to do with other subjects, to inquire into the relevant facts. —J. L. Brierly, *The Outlook for International Law* (Oxford: The Clarendon Press, 1944), 1–2.

This book constitutes an inquiry into the "relevant facts" concerning whether environmental treaties influence international behavior and by what methods they succeed in doing so. The introductory section acquaints the reader with the substantive and theoretical foundations that the balance of the book uses in empirically evaluating whether and how international efforts to regulate intentional oil pollution have caused behavioral change. This section's three chapters describe the current status of international environmental law, noting the growing number of environmental treaties and the increasing concern regarding the need to successfully induce compliance. They set out the methodology to be used in the empirical research and frame the current debate between skeptics and believers in the influence of international law. The section concludes by describing the historical background of the treaties and amendments that have been negotiated to control intentional discharges of oil by tankers.

1
Environmental Treaty Compliance: An Introduction

Treaty rules matter. When facing unsatisfactory levels of compliance, negotiators can modify environmental treaties to increase compliance. Nations can restructure treaty compliance systems to increase compliance, even in the absence of greater environmental concern. Although the level of environmental concern and the power of those states seeking international action set bounds on compliance levels, the rights and obligations that treaty rules establish for compliance, monitoring, and enforcement determine how successful the treaty will be at changing existing patterns of behavior. Some rules work better than others. The degree to which governments and private actors adapt their behavior to conform with environmental treaty provisions depends upon the regulatory strategy underlying those provisions, how the provisions are framed, and the structure of the attendant compliance system for preventing, monitoring, and responding to violations.

International treaties and regimes have value if, and only if, they cause people to do things that they would not otherwise do. Indeed, "the object of the agreement is to affect state behavior."[1] Yet anecdotal evidence suggests that nations regularly violate even relatively low-cost reporting requirements, and often ignore more substantive provisions.[2] The fact that compliance is crucial to treaty success but cannot be automatically

1. Abram Chayes and Antonia Handler Chayes, "On Compliance," *International Organization* 47 (Spring 1993), 193.

2. See, for example, United States General Accounting Office, *International Environment: International Agreements Are Not Well Monitored* (Washington, DC: GPO, 1992); and Gerard Peet, *Operational Discharges From Ships: An Evaluation of the Discharge Provisions of the MARPOL Convention by Its Contracting Parties* (Amsterdam: AIDEnvironment, 1992).

assumed provides the foundation for the major questions raised by this book. Most fundamentally, it asks, Can treaties induce behavioral change?[3] Do treaty language, procedures, and processes play any role in determining the level of compliance that a treaty elicits? Or is behavioral change overwhelmingly determined by other factors, in which case efforts to "improve" treaty compliance are misguided? If treaty rules and compliance systems do make a difference—if some treaties do elicit greater compliance than others, controlling for other factors—what features explain success and failure? If self-reporting is as common and as crucial to environmental treaties as is often claimed, what types of systems encourage it? Do compliant states always report? Do noncompliant states always fail to report? Given the obstacles to sanctioning even when crucial security and economic interests are at stake, how can states successfully deter those who intentionally violate environmental accords? Do governments use the promise of their own compliance and the threat of violation to induce compliance by recalcitrant states? How do we draft new environmental accords that elicit higher levels of compliance than past ones? What does experience teach us about the types of rules that produce compliance? If Maurice Strong, chairman of the United Nations Conference on Environment and Development (UNCED), is right that "we have to push like hell to make sure implementation takes place," then in which direction should we push?[4]

The Debate over Treaty Influence

It may seem odd to ask, Can treaty rules induce behavioral change? To many, the answer may appear so obvious that it does not warrant spending the time to research the matter. The obvious answer, however,

3. Although the questions addressed in this book grow out of, are informed by, and are intended to contribute to the larger debate on the influence of regimes, the book is limited to a discussion of explicit rules delineated in treaties and the attendant processes and procedures established to elicit compliance. I consciously do not address the norms and principles included in the standard definition of regimes. For such a definition, see Stephen D. Krasner, ed., *International Regimes* (Ithaca: Cornell University Press, 1983), 2.

4. Maurice Strong, quoted in James Brooke, "U.N. Chief Closes Summit with an Appeal for Action," *New York Times*, 15 June 1992, A8.

is yes for some and no for others. On the "yes" side, international diplomats have signed treaties regulating a wide array of pollutants, from oil and industrial wastes that are dumped in the oceans to nuclear fallout to river pollutants and acid precipitation. They have spent years negotiating and refining accords to protect seals, whales, fisheries, polar bears, endangered species, wetlands, tropical timbers, and world heritage sites.[5] Negotiators spend considerable time formulating and reformulating treaty language not only to reach agreement, but often with the hope that the final draft will prove more successful at eliciting compliance than earlier versions. National representatives' actions and rhetoric reflect a belief that international commitments constrain their future options. Governments' frequent calls for stricter treaty provisions, like the European pressures at UNCED to establish targets and timetables for reducing greenhouse gas emissions, rest on the premise that stronger language will produce greater changes in behavior. Other nations' attempts to weaken the terms of an agreement are equally founded in the belief that, once adopted, the treaty will limit their policy options. Even after final negotiation, countries frequently refuse to sign an international environmental agreement—despite the significant political benefits of appearing "green"—precisely because they perceive the treaty as imposing legally binding constraints on behavior, constraints they can avoid

5. Bernd Ruster and Bruno Simma's thirty-odd-volume set entitled *International Protection of the Environment: Treaties and Related Documents* (Dobbs Ferry, NY: Oceana Publications, 1975) is the most comprehensive compilation of environmental treaty texts and proves invaluable for primary research into international environmental treaties. For modern legal texts, see Edith Brown Weiss, Daniel Barstow Magraw, and Paul C. Szasz, *International Environmental Law: Basic Instruments and References* (Dobbs Ferry, NY: Transnational Publishers, 1992). Michael Molitor, *International Environmental Law: Primary Materials* (Boston: Kluwer Law and Taxation Publishers, 1991) provides the texts of more notable and recent environmental treaties. The U.N. Environmental Program's *Register of International Treaties and Other Agreements in the Field of the Environment* (Nairobi, Kenya: United Nations Environment Program, 1989) provides brief descriptions of most major multilateral agreements. For more detailed analyses, see Simon Lyster, *International Wildlife Law: An Analysis of International Treaties Concerned with the Conservation of Wildlife* (Cambridge, England: Grotius Publications, 1985); and Peter H. Sand, *Marine Environment Law in the United Nations Environment Programme: An Emergent Eco-regime* (London: Tycooly, 1988).

by not signing. In some cases, environmental treaties actually appear to have made major contributions to solving problems, as in the case of fur seal treaties' bringing back populations that were in sharp decline.[6]

Nongovernmental actors likewise act as if treaty rules matter. Corporations expend considerable effort and resources to lobby both domestically and internationally against, and sometimes for, environmental treaties. For example, many American genetic research and drug development companies opposed U.S. signature of the biodiversity treaty because they believed that, under the treaty, "our hands would be tied" because it would "force them to share patents, profits, and technology with developing countries, giving those countries an unfair competitive advantage."[7] Environmental nongovernmental organizations (NGOs) also spend considerable resources lobbying for or against specific treaty provisions, and frequently press for stronger regulations, the major efforts made to establish a whaling moratorium being a case in point.[8] Even the frequent calls for "treaties with teeth" simultaneously attest to a belief that treaties can work, even if they often do not.

On the "no" side of the question of whether treaty rules induce behavioral change, many people view treaties more cynically. This perspective is epitomized by the notion that treaties reflect merely the lowest common denominator of the preferences of the states negotiating them.[9] In this view, treaties merely codify the behavior that the least activist, most reluctant nation will accept. Member nations negotiate and sign precisely those treaties that reflect their preexisting interests in engaging in or refraining from the behaviors specified in the treaties. For example,

6. *Interim Convention on the Conservation of North Pacific Fur Seals*, 9 February 1957, T.I.A.S. no. 3948, 314 U.N.T.S. 105, 8 U.S.T. 441, reprinted in 11 I.L.M. 251. Simon Lyster discusses these treaties in *International Wildlife Law*, 47ff.

7. Ronald Rosenberg, "Industry Calls Biodiversity Pact Unfair," *Boston Globe*, 12 June 1992, 9.

8. M. J. Peterson, "Whalers, Cetologists, Environmentalists and the International Management of Whaling," *International Organization* 46 (Winter 1992).

9. See, for example, Peter Sand's "International Cooperation: The Environmental Experience," in Jessica Tuchman Mathews, ed., *Preserving the Global Environment: The Challenge of Shared Leadership* (New York: W. W. Norton and Co., 1991).

nations appear likely to phase out ozone-depleting substances even before required by the Montreal Protocol.[10] For years, environmentalists accused the International Whaling Commission of being a "whaling club" in which violations were rare because the commission never set quotas below the levels of harvest desired by the whaling nations.[11]

People on this side of the question feel that treaties reflect power and interests, but do not shape behavior. When behavior conforms to treaty rules, it does so because both the behavior and the rules reflect the interests of powerful states. Compliance is an artifact of one of three situations:

1. the treaty rules merely codify the parties' existing behavior or expected future behavior,
2. a hegemonic state enforces the treaty or induces other actors to comply, or
3. the treaty rules govern a coordination game wherein parties have no incentives to violate once a stable equilibrium has been achieved.[12]

Treaty rules correlate with, but do not cause, compliance. Therefore, efforts to improve treaty rules to increase compliance reflect either the changed interests of powerful states or are misguided exercises in futility. Compliance is a function of factors not amenable to manipulation by international policymakers or treaty drafters. In this realist conception, whenever nations have incentives to violate treaty proscriptions they will do so, as was evidenced by Russia's dumping radioactive waste at sea ten

---

10. *Vienna Convention for the Protection of the Ozone Layer*, 22 March 1985, T.I.A.S. no. 11097, reprinted in 26 I.L.M. 1529 (1987), hereinafter cited as *Vienna Convention;* and the *Montreal Protocol on Substances That Deplete the Ozone Layer*, 16 September 1987, reprinted in 26 I.L.M. 1541 (1987), hereinafter cited as *Montreal Protocol.*

11. *International Convention for the Regulation of Whaling*, 2 December 1946, T.I.A.S. no. 1849, 161 U.N.T.S. 72. See any of the International Whaling Commission's annual *Report of the International Whaling Commission* (London: The International Whaling Commission).

12. Arthur A. Stein, "Coordination and Collaboration: Regimes in an Anarchic World," in Krasner, ed., *International Regimes.* For a more extended version of this distinction, see Arthur A. Stein, *Why Nations Cooperate: Circumstance and Choice in International Relations* (Ithaca: Cornell University Press, 1990).

years after a moratorium on such dumping took effect.[13] In 1979 and 1980 alone, three hundred violations of the convention restricting trade in endangered species came to light.[14] While European nations agreed to reduce sulfur dioxide emissions by 1993, current emissions make significant noncompliance likely.[15] Some treaties protecting wildlife and their habitats have become "sleeping conventions," ignored or forgotten by signatories.[16] As Winston Churchill claimed, nations will only "keep their bargains as long as it is in their interest to do so."[17]

Despite the obvious differences between these outlooks, they lead to similar predictions regarding behavior. Both views suggest that nations and their corporate and private citizens will frequently fulfill treaty commitments. International lawyers like Louis Henkin claim that "almost all nations observe almost all principles of international law and almost all of their obligations almost all of the time," while the father of modern realism, Hans Morgenthau, claims that "the great majority of the rules of international law are generally observed by all nations without actual compulsion."[18] Both views also expect violations to oc-

13. David E. Sanger, "Nuclear Material Dumped Off Japan," *New York Times* (October 19, 1993), A1.

14. *Convention on International Trade in Endangered Species of Wild Fauna and Flora*, 3 March 1973, T.I.A.S. no. 8249, 27 U.S.T. 1087, 983 U.N.T.S. 243, reprinted in 12 I.L.M. 1085. Lyster, *International Wildlife Law*, 271.

15. Marc Levy, "European Acid Rain: The Power of Tote-board Diplomacy" in Peter Hass, Robert O. Keohane, and Marc Levy, eds., *Institutions for the Earth: Sources of Effective International Environmental Protection* (Cambridge, MA: The MIT Press, 1993).

16. See, for example, the *African Convention on the Conservation of Nature and Natural Resources*, 15 September 1968, 1001 U.N.T.S. 3, reprinted in 5 I.P.E. 2037; and the *Convention on Nature Protection and Wildlife Preservation in the Western Hemisphere*, 12 October 1940, U.S.T.S. 981, 161 U.N.T.S. 193, reprinted in 4 I.P.E. 1729. Simon Lyster analyzes these in "Effectiveness of International Regimes Dealing with Biological Diversity, Including Wildlife Conservation and Habitat Protection, from the Perspective of the North or Developed World," unpublished paper, Godalming, England, June 1991, 2.

17. Cited in Hans Morgenthau, *Politics Among Nations: The Struggle for Power and Peace* (New York: Alfred A. Knopf, 1978), 560.

18. Louis Henkin, *How Nations Behave: Law and Foreign Policy* (New York: Columbia University Press, 1979), 47; and Morgenthau, *Politics Among Nations*, 299. Oran Young claims that "states generally comply with the rights and rules of international institutions"; see *International Cooperation: Building Regimes*

cur: the former claims only that treaties can effect behavior, not that they always will, and the latter claims that violations merely reflect individual cases of conflict between treaty requirements and immediate interests. As with many issues, the debate is not over the facts, but over how to interpret the facts. As Chayes and Chayes have noted, the debate often devolves into a question of which assumptions one adopts.[19]

To resolve this debate requires that one surmount several difficulties. Even the greatest proponents of treaties do not contend that all treaty rules elicit compliance. Therefore, identifying instances of treaties that failed to influence behavior does not falsify the claim that treaties can elicit compliance but only clarifies that they do not always do so. In contrast, those who argue that treaties have little impact on behavior do not contend that compliance is infrequent. They argue only that treaty rules and compliance are not causally linked but are separate yet correlated indicators of underlying interests and incentives over which a treaty has no influence. Thus high compliance levels and even changes in compliance that correlate with treaty change prove insufficient to convince a naysayer that treaty rules matter. One must not only show a strong correlation and a strong causal link between adoption of certain rules and compliance, but must also show the absence of other equally plausible causal links.

Even those who argue that treaty rules matter often contend that new rules and behavior changes to comply with them are not tightly linked by clear and immediate causal pathways. Richard Benedick has claimed that UNCED "should not be judged by the immediate results, but by the process it sets in motion," and that the results might not be apparent for years.[20] The influences of treaty rules on behavior take time, with a multitude of other factors contributing to, and intervening in, the process. These causal pathways are often indirect, attenuated, and not readily subject to empirical, nonsubjective verification, raising obstacles to efforts to convincingly argue, let alone prove, treaty impact.[21] If these are

---

for Natural Resources and the Environment (Ithaca: Cornell University Press, 1989), 62.

19. Chayes and Chayes, "On Compliance."

20. William K. Stevens, "Lessons of Rio," *New York Times*, 14 June 1992, 10.

21. Chayes and Chayes, "On Compliance."

the only pathways of influence, then empirical evidence that treaties can influence behavior will prove truly hard to find.

The Argument and Central Findings

This book starts with the premise that empirical research can move this debate beyond unsupported assumptions and assertions to clearly identify whether treaties can influence international behavior in environmental affairs and, if so, the conditions under which they can. This book's analysis of almost forty years of experience with treaties regulating intentional oil pollution demonstrates that treaty rules can positively influence international behavior at both governmental and corporate levels. The argument consists of four points. First, there is considerable noncompliance with many treaty provisions, bringing into question both sides' assumptions that international law is generally observed, at least in environmental affairs. Second, this noncompliance is not ubiquitous; there is extraordinary variance between levels of compliance across the provisions of a single treaty. In three cases that are described, different regulatory strategies that targeted the same underlying behavior produced strikingly different levels of compliance. Third, a major share of this variance remains even after one has controlled for the effects of power, interests, and other factors exogenous to the treaty. The difference in compliance levels exists even when the same countries are compared over the same time period. Fourth, the success of some rules and the failure of others to elicit compliance—the remaining variance—can be explained only by reference to the character of the treaty rules.

Even a single case in which there is clear and convincing evidence of compliant behavior that would not have occurred in the treaty's absence provides an "existence proof" that treaty rules can influence behavior.[22] As Robert Keohane has noted, realism's "taut logical structure and its pessimistic assumptions about individual and state behavior serve as barriers against wishful thinking" regarding the value of treaty making.[23] While avoiding the Scylla of excessive faith that treaties, regimes, and

22. Abram Chayes, personal communication.

23. Robert O. Keohane, *After Hegemony: Cooperation and Discord in the World Political Economy* (Princeton, NJ: Princeton University Press, 1984), 245.

institutions will produce their intended results, however, we must also avoid the Charybdis of a too-easy dismissal of the idea that they can have some influence on behavior. This book argues that it is possible to do both. It documents several cases in which treaty provisions and compliance systems can be identified as the cause of behavioral change.

The book provides strong evidence that improvements in treaty compliance can be achieved without waiting for independent forces to alter the perceived benefits of compliance or for an environmentally conscious state to gain sufficient power to bring other states in line. Even without such changes, treaty provisions can be refined in ways that make compliance more likely. Both the theoretical and the empirical chapters develop the concept that compliance systems are made up of three major components: the primary rule system, the compliance information system, and the noncompliance response system.[24] Treaty compliance systems succeed by ensuring that the three tasks of compliance, monitoring, and enforcement are defined in ways that place the burden of performing those tasks on actors that have the political and economic incentives, the practical ability, and the legal authority to carry them out. While exogenously determined power and interests establish the context in which a treaty must operate, this context underdetermines the observed levels of compliance. Treaty drafters often have considerable latitude to select among alternative proscriptions and prescriptions, as well as among supporting measures for monitoring and responding to violations. Their choices play a crucial role in determining how likely compliance will be. They can make compliance quite likely by seizing opportunities to place actors playing all three roles within an incentive-ability-authority triangle, a strategic triangle of compliance.[25] The governments of states that opposed certain oil pollution regulations have monitored and enforced them and reported on their enforcement, and their oil and shipping companies have complied even when facing strong economic disincentives to doing so.

The cases described in this book suggest specific propositions regarding the types of compliance systems that work. One proposition is that

24. These terms are defined in chapter 2.

25. I am indebted to Robert O. Keohane for this notion of a strategic triangle of compliance.

"opportunistic" primary rule systems increase compliance more by matching burdens and responsibilities with actors' interests and capabilities than by attempting to alter those interests and capabilities. The evidence suggests that two rules, both of which are acceptable given member states' existing power and interests, may have quite different implications for the level of compliance achieved. Whether trying to elicit compliance, reporting, monitoring, or enforcement, successful rules have facilitated and removed barriers to desirable action by those actors with existing interests in taking those actions. For example, when tanker operators demonstrated an ongoing willingness to intentionally discharge oil in violation of international rules, these discharges were reduced by requiring owners to buy tankers that essentially prevented such discharges. The nature of the tanker procurement process and the incentives of the actors involved made tanker owners much more compliant than tanker operators despite the significantly higher costs involved for the former.

Another proposition is that effectively collecting information on compliance depends upon establishing a system that provides direct benefits to the information providers. Whether they involve self-reporting or independent verification, the compliance information systems that have succeeded in eliciting compliance data have done so by making use of the information collected to further the interests of those providing it. In a self-reporting system, not only must sanctioning of reported noncompliance be precluded, but the system must collect, analyze, and disseminate the information collected in ways that provide the reporters with otherwise unavailable benefits. In a system that relies on independent verification, those providing evidence of noncompliance must be protected from retaliation and reported noncompliance must elicit some type of response that makes future compliance more likely. Compliance information systems also benefit from building on existing information infrastructures rather than creating new ones. Systems monitoring tanker compliance and those monitoring governmental enforcement activity have both succeeded when they have built on preexisting information infrastructures. Compliance information systems succeed by analyzing and disseminating the information provided, facilitating and standardizing reports, and linking themselves to bureaucratic standard operating procedures.

Another proposition suggested by the cases developed in this book is that existing legal and practical barriers constrain the responses of actors that view themselves as harmed by noncompliance. Rules that remove such obstacles or impediments increase sanctioning far more than those that impose obligations for states to use sanctions that they are already capable of using and authorized to use. Effective responses, in turn, must be both appropriate and adequate to remedy the cause of the noncompliance. Noncompliance arising from a lack of financial resources will not be remedied by sanctions or negative publicity. However, even when sanctions are appropriate, treaty rules must recognize that sanctions proportional to the harm of a violation may well not prove adequate to deter future violations by others. When neither flag states nor oil-loading states have proved willing to engage in rigorous penalization of discharge standards, treaty rules have improved enforcement by allowing developed states to detain ships violating equipment regulations without requiring the permission of the flag state. This policy has simultaneously shifted the right to respond to those with incentives to respond while providing them with a response that was adequate to deter violations.

The evidence also strongly supports a view of compliance with environmental treaties as a two-level enforcement game that mirrors the two-level negotiation game developed by Putnam.[26] The book demonstrates that governments often have only imperfect control over environmentally harmful behaviors. Treaties addressing arms control, human rights, or international banking and trade predominantly, if not exclusively, target government actions for regulation. In contrast, the primary, if often indirect, targets of much environmental regulation are individual and corporate actions not directly within governmental control. To effectively elicit behavioral change requires environmentally concerned states either to find mechanisms for directly influencing the behavior of polluters who are nationals of other countries or to find means to induce reluctant governments to influence that behavior themselves. Environmental treaties are prone to these problems that are not evident in most other issue areas.

26. Robert D. Putnam, "Diplomacy and Domestic Politics: The Logic of Two-Level Games," *International Organization* 42 (Summer 1988).

A final important finding of this book is that reciprocity—the use of retaliatory noncompliance to elicit compliance—is rarely used despite the fact that significant noncompliance by both governments and industry is common. Nonreporting also proves to be quite common among several of the reporting systems established, although the evidence clearly demonstrates that nonreporting cannot be assumed to equate to noncompliance. Many developed states meeting a treaty's substantive provisions nonetheless regularly fail to provide reports. The reporting systems also point to the major obstacles that prevent secretariats from disseminating compliance information, despite the conventional wisdom that such dissemination is a crucial mechanism for eliciting greater compliance.

Existing Literature

The desire to answer questions about international environmental politics, especially about compliance, increased in the early 1990s among both policymakers and academics. In the policy community, this concern became manifest with the creation of the Commission on Sustainable Development in the wake of UNCED, specifically to address the implementation of Agenda 21, an extensive international blueprint for environmental action. In academic circles, the interest is evident in a plethora of current and recently completed research efforts throughout the United States and Europe.[27]

This growing interest in international environmental affairs in general, and in compliance in particular, can be traced to the end of the Cold War, which prompted a major refocusing of policy and scholarly interest

27. These include, among others, book manuscripts being written and projects being conducted at or with funding from Dartmouth College (Oran Young and Marc Levy); the European Science Foundation (Kenneth Hanf); the Foundation for International Environment Law and Diplomacy (James Cameron); the Fridtjof Nansen Institute (Steinar Andresen); Harvard University (Abram Chayes and Antonia Chayes); Harvard University (William Clark, Robert Keohane, and Marc Levy); the International Institute for Applied Systems Analysis (David Victor and Eugene Skolnikoff); the Social Science Research Council (Edith Brown Weiss and Harold Jacobson); and the University of Tubingen (Volker Rittberger). One published analysis is Haas, Keohane, and Levy, *Institutions for the Earth*.

away from security and onto other international concerns, including the environment. A simultaneous increase in the perceived number and gravity of global environmental risks produced treaties addressing such issues as climate change, biodiversity loss, and stratospheric ozone depletion. These were only the most recent attempts to use international law to alter the governmental, corporate, and individual activities that threaten the global environment. The trend certainly suggests that nations and their citizens will continue to turn to treaties as the strategy of choice for managing transboundary and global environmental problems. This recent wave of environmental treaties coincided with a period during which the earlier wave of treaties had accumulated sufficient evidence to suggest that compliance does not always follow treaty ratification and to provide a basis for drawing conclusions that could enhance the prospects for future compliance.[28]

Previous academic interest in international environmental affairs had produced solid descriptions of various environmental regimes and the processes of treaty and regime development, but had provided only limited evaluations of their effectiveness.[29] Other works have focused on environmental regime formation rather than upon efforts to implement treaties and improve compliance over time.[30] This trend has continued, especially in the wake of the Montreal Protocol, with numerous authors seeking to examine and explain the conditions that lead nations to negotiate environmental agreements.[31] Prescriptive proposals regarding

28. See, for example, Sand, "International Cooperation."

29. Lynton Keith Caldwell, *International Environmental Policy* (Durham, NC: Duke University Press, 1984); John E. Carroll, ed., *International Environmental Diplomacy: The Management and Resolution of Transfrontier Environmental Problems* (Cambridge, England: Cambridge University Press, 1988); and David A. Kay and Harold K. Jacobson, eds., *Environmental Protection: The International Dimension* (Totowa, NJ: Allanheld, Osmun & Co., 1983).

30. See, for example, Peter M. Haas, *Saving the Mediterranean: The Politics of International Environmental Cooperation* (New York: Columbia University Press, 1990); Young, *International Cooperation*; and Oran R. Young and Gail Osherenko, eds., *Polar Politics: Creating International Environmental Regimes* (Ithaca: Cornell University Press, 1993).

31. See, for example, Sharon L. Roan, *Ozone Crisis: The 15 Year Evolution of a Sudden Global Emergency* (New York: John Wiley, 1989); and Richard Elliot Benedick, *Ozone Diplomacy: New Directions in Safeguarding the Planet* (Cambridge, MA: Harvard University Press, 1991).

the form and substance of climate change, deforestation, and biodiversity treaties have also abounded.[32]

Only recently have any authors made concerted efforts to evaluate compliance and effectiveness. Young and Fisher wrote seminal works on treaty compliance theory over a decade ago, but further work in this area languished.[33] "Up to now the international regime approach has been largely hypothetical in nature with respect to environmental affairs," and rigorous empirical identification of the causes of compliance only recently became a focus of study.[34] While many of the works already mentioned have addressed compliance issues, they have not made it their central focus, nor have they limited their inquiries to the role of treaties in influencing behavior.[35] One volume examines the effectiveness of envi-

32. For some of the better examples addressing climate change, see James K. Sebenius, "Designing Negotiations toward a New Regime: The Case of Global Warming," *International Security* 15 (Spring 1991), and Michael Grubb, "The Greenhouse Effect: Negotiating Targets," *International Affairs* 66 (January 1990). On deforestation, see Kenton R. Miller, Walter V. Reid, and Charles V. Barber, "Deforestation and Species Loss: Responding to the Crisis," in Jessica Tuchman Mathews, ed., *Preserving the Global Environment: The Challenge of Shared Leadership* (New York: W. W. Norton and Co., 1991). On biodiversity, see Edward O. Wilson, "The Biological Diversity Crisis: A Challenge to Science," *Issues in Science and Technology* 2 (Fall 1985).

33. Oran Young, *Compliance and Public Authority: A Theory with International Applications* (Baltimore: Johns Hopkins University Press, 1979); and Roger Fisher, *Improving Compliance with International Law* (Charlottesville, VA: University Press of Virginia, 1981). For recent efforts to remedy this, see Harold Jacobson and Edith Brown Weiss, "Implementing and Complying with International Environmental Accords: A Framework for Research," paper presented at the annual meeting of the American Political Science Association, San Francisco, CA, 30 August - 2 September 1990; Chayes and Chayes, "On Compliance"; Peter H. Sand, "International Cooperation"; and Philipp M. Hildebrand, "Towards a Theory of Compliance in International Environmental Politics," paper presented at the annual meeting of the International Studies Association, March 1992.

34. Volker Prittwitz, "Several Approaches to the Analysis of International Environmental Policy," paper presented at a conference entitled "What Price Environment?" at the Swedish Institute of International Affairs, June 1988, 21.

35. In *Pollution, Politics, and International Law: Tankers at Sea* (Berkeley: University of California Press, 1979), R. Michael M'Gonigle and Mark W. Zacher provide a phenomenally rich and detailed analysis of compliance and enforcement as one part of their broader analysis of international oil pollution regulations. They were prevented from analyzing the impact of the most sig-

ronmental regimes, but includes several nontreaty regimes and does not limit its focus to compliance and behavioral change.[36] Trexler provides an in-depth empirical evaluation of the impact of the Convention on International Trade in Endangered Species, but does not use a methodology that isolates the impact of the treaty from other factors.[37]

This book also builds on efforts to examine regimes in other areas. Here too, however, most research has focused on identifying the sources of cooperation and explaining regime formation rather than on rigorously exploring and testing theories regarding compliance and regime effectiveness. Years after Krasner's edited volume appeared, authors were still asking the question "Do regimes matter?"[38] Recently researchers have begun to dedicate serious resources to investigating the components and determinants of regime effectiveness in such areas as alliance force structuring and international finance.[39] Fortunately much of this work appears to be focusing on environmental regimes, and it certainly will provide valuable insights into the role of treaty rules and processes in determining compliance. This book seeks to contribute to that enterprise by helping to fill the significant gap in empirically based understanding of the role that international agreements play in determining the environmental behavior of nations and their citizens. It takes up Oran Young's challenge for "careful empirical work . . . comparing and

---

nificant changes in these regulations, which did not enter into force until four years after the publication of their book. Peter M. Haas also addresses compliance issues, but not as the central focus of his *Saving the Mediterranean*.

36. Haas, Keohane, and Levy, eds., *Institutions for the Earth*.

37. Mark C. Trexler, "The Convention on International Trade in Endangered Species of Wild Fauna and Flora: Political or Conservation Success?" unpublished Ph.D. dissertation, University of California at Berkeley, Berkeley, CA, December 1989.

38. See, for example, Peter M. Haas, "Do Regimes Matter? Epistemic Communities and Mediterranean Pollution Control," *International Organization* 43 (Summer 1989); Stephen Haggard and Beth A. Simmons, "Theories of International Regimes," *International Organization* 41 (Summer 1987); and Young, *International Cooperation*, 206.

39. Ethan Kapstein, *Governing the Global Economy: International Finance and the State* (Cambridge, MA: Harvard University Press, 1994), and John S. Duffield, "International Regimes and Alliance Behavior: Explaining NATO Conventional Force Levels," *International Organization* 46 (Autumn 1992).

contrasting real-world regimes. . . . to isolate the role of institutional arrangements in shaping collective behavior."[40]

Analytic Goals, Methods, and Limitations

This book contributes to these research programs by providing a rigorous empirical analysis of treaty rules and compliance. Indeed its major goal is to develop extensive documentation that goes beyond a compilation of anecdotal evidence, producing more rigorous and reliable evidence of behavioral change that can be unambiguously attributed to a treaty. The book guards against the Scylla of unexamined faith in treaties mentioned above through careful case selection and the use of rigorous analytic procedures.

Since the study's goal is to identify the extent of treaty impact on behavior, I take the realist skepticism seriously and start by assuming that changes in environmental treaty rules cannot cause an independent increase in compliance. The methodological goal then becomes to identify and structure the analysis so that the empirical evidence will be of the type needed to falsify this assumption. Whether the analysis will be able to falsify the assumption depends on the evidence, but the methodological structure ensures that the evidence developed addresses the realist skepticism.

The study uses treaty rules as the units of analysis. I examine the relationship between changes in rules and changes in compliance within a given treaty rather than examining the differences in rules and differences in compliance across different treaties. Rather than being a study of a single case of compliance with intentional oil pollution regulation, this is a study of several cases of rule change within the context of intentional oil pollution regulation. Analysis at this level has two advantages. First, it is analytically more accurate than analysis of a single case since, as is discussed in chapter 2, actors do not comply with treaties, but with specific treaty provisions. Second, by analyzing several rules in a single treaty, factors that cause variance in compliance across issue areas can be held constant and these factors can be eliminated as possible causes of variance, reducing interpretive ambiguity and making "it pos-

40. Young, *International Cooperation*, 209.

sible to attribute variance in collective outcomes to the impact of institutional arrangements with some degree of confidence."[41]

I selected the 1954 International Convention for the Prevention of Pollution of the Sea by Oil (OILPOL) and its successor, the 1973 International Convention for the Prevention of Pollution from Ships (known as MARPOL) as the treaty context for several reasons. The OILPOL treaty experienced low levels of initial compliance. High initial compliance levels would raise questions about whether actors had any incentives to violate and would suggest that the treaty was addressing a coordination or harmonization problem, in which case incentives for violation would be insignificant. Treaties with high levels of compliance from the beginning, like the Limited Test Ban Treaty or the Antarctic Treaty, are unlikely to represent collaboration problems, compliance with which is most difficult to explain.[42] Continuing noncompliance and efforts to improve enforcement show that the OILPOL and MARPOL rules were not self-enforcing, confirming the characterization of oil pollution as a collaboration problem.

Comparing rules within a single treaty context also required a case in which concerns over noncompliance produced changes in the treaty compliance system. The rule changes had to have been made long enough after initial treaty signature to allow for the establishment of baseline compliance levels and long enough ago to identify any changes in compliance levels after the treaty change. Amendments to OILPOL and MARPOL have been made throughout the forty-year history of ocean oil pollution regulation.

Additionally, reasonably good data had to be available on compliance and noncompliance as well as on nontreaty factors that would significantly affect the incentives to comply with or violate the treaty rules. Treaties with exclusively hortatory requirements would have made data on compliance too dependent upon subjective evaluations by the analyst. Although working within a single treaty context held many variables constant, the oil pollution issue also involves a market in which any shifts

41. Young, *International Cooperation*, 208.

42. For example, Young devotes a full chapter of *Compliance and Public Authority* to the Limited Test Ban Treaty, which has had negligible violations.

in incentives and power could be identified through an analysis of data on proxy variables such as oil prices, oil imports, and tanker registries.

Finally, intentional oil pollution from tankers poses a particularly hard case for analysis, i.e., theory predicts that treaty rules are very unlikely to result in compliance. Collective action theory predicts that it will prove quite difficult to force adoption of, let alone compliance with, international rules that require the powerful and concentrated oil industry to incur large pollution control costs to benefit the public at large.[43] The costs of compliance to the industry have also increased over time, suggesting that compliance with the increasingly stringent standards should decrease over time.

Within the oil pollution context, I selected individual cases that provided variance across rules while holding most other factors constant. These included cases involving rule change over time and cases comparing two rules during the same period of time. The major analytic challenge in studying the former cases is to identify exogenous factors that changed at the same time as a rule changed and assess whether the exogenous factors or the rule change more plausibly explain any observed increase in compliance. For example, the replacement of initial 1954 discharge standards with more specific and verifiable discharge standards in 1969 allows for an evaluation of the same behavior—tanker discharges—before and after adoption of the new rules. Performing such a time series analysis allows one to avoid the risk of confusing the direction of causality by looking for change in compliance *after* change in a rule. It also allows the researcher to hold constant many variables that might change across behaviors.[44] All exogenous factors that did not change during the same time period as the rule can be eliminated as explanations of why compliance levels changed.

Studying cases that allow one to compare two different rules regulating similar behavior over the same time period allows for better control of most variables, but such cases are less common because such regulatory

43. See, for example, Mancur Olson, *The Logic of Collective Action: Public Goods and the Theory of Groups* (Cambridge, MA: Harvard University Press, 1965), 34.

44. In short, it reduces the likelihood of multicollinearity problems' confounding the analysis; see James D. Fearon, "Counterfactuals and Hypothesis Testing in Political Science," *World Politics* 43 (January 1991), 186.

overlap is rare. The major analytic challenge in such cases is to ensure that the regulated behaviors are sufficiently similar that the exogenous incentives to undertake them are identical and only the difference between the two rules can account for the variance in behavioral response. For example, during the 1980s European states had to report similar types of information on enforcement through quite different systems run by the International Maritime Organization and the Memorandum of Understanding on Port State Control.

After identifying cases of these types, I follow six analytic steps to determine whether and how treaty rules influenced behavior. First, I identify the exact changes in the independent variable of interest, that is, treaty rules that were adopted in an effort to increase compliance. This provides the crucial variance in the policies that is the focus of the study. Then I investigate whether there were changes in compliance subsequent to the rule change or differences in compliance across rules. Measuring the dependent variable—compliance—poses major theoretical challenges. There are obvious obstacles to determining actual rather than detected violations: violations may go undetected, and detected ones may often go unreported or reports may not be disseminated. As subsequent chapters show, the absence of data cannot be safely interpreted as evidence of either compliance or violation. Aggregating compliance data across actors and over time poses many problems. The MARPOL convention required some tankers to retrofit segregated ballast tanks after a specified year. Should ongoing violations—failure to retrofit until five years later—count as a single violation or as multiple violations? If all of one country's tankers fail to retrofit while only five of one hundred in another country fail to do so, should both governments be counted as in noncompliance, or should we consider the second country as 95 percent in compliance? Should a tanker that retrofits after the deadline be considered as complying? Nondichotomous activities, such as discharging oil, raise even greater potential difficulties. MARPOL prohibits discharges in excess of sixty liters per mile. Should tankers that discharge sixty-one liters per mile be aggregated with those that discharge five hundred liters per mile? Should a tanker that discharged one hundred liters per mile for five miles on one voyage in 1984 be counted as having fewer or more violations than one that made half-mile discharges of one hundred liters per mile on five voyages? One would also like to know a violation rate, reflecting the number of cases in which actors violated as a proportion of the cases in which those actors had opportunities to violate. But how does one identify "opportunities to violate"? Does compliance increase when all actors comply more often, or when the "average" actor complies more often? What happens when more states sign a treaty?

These and related questions demand more attention. Fortunately, perhaps, hobbling data limitations prevent even the possibility of measuring compliance in the ways theory might suggest. As is true in many research efforts, practical data problems have made theoretical concerns moot. This book makes the best use of available quantitative and qualitative data to address the theoretical concerns above. I use time series data from the same source wherever possible to keep data consistent over time. I also use raw compliance and violation data and also, where possible, normalize the data using proxies that control for changes in the activity being regulated. I exploit independent sources to cross-check my assessments of trends in compliance. I deal with the many remaining data problems by requiring myself to conclude that treaty rules have had no impact unless clear evidence demonstrates that compliance has increased and that factors other than the treaty cannot account for that increase.

In cases in which compliance has not changed, I look for features of the rule change that might explain why it failed to elicit greater compliance. If actors continue business as usual, after a change in rules, then international rules are clearly ineffective. The failure of a treaty rule to influence compliance may nonetheless prove fruitful for analysis. Failures are of little interest to the realist, since they are to be expected. However, if evidence shows that one rule has increased compliance while another has not, then we need to evaluate what features distinguish the failures from the successes.

If actors behave differently under one set of rules and other factors likely to have induced such a behavioral change cannot be found, then the rules can be assessed as being effective. In cases in which compliance has changed, I develop a causal story, including supporting empirical evidence, demonstrating how the rule change could have led to the observed change in behavior. I identify the pathway by which the rule altered behavior and provide evidence that the treaty rule, rather than other factors, caused the chain of events involved.

Next, I evaluate rival explanations of how exogenous factors can explain the observed change in behavior, evaluating whether any variance that occurred in such factors would be predicted to have increased compliance. Careful case selection allowed me to hold many variables constant, excluding from consideration variables that experienced no change, but we must look for the ways and evidence that treaty-independent economic and technological changes or domestic political changes in key countries can more readily explain observed compliance changes. Since the realist claim that compliance changes and rule changes are both indicators of shifts in underlying power and interests can verge on tautology, I develop proxy variables other than the changes in rules themselves to identify changes in the power and incentives of actors to violate treaty rules, for example, oil imports and oil prices.

Finally, I use counterfactuals to assess what the levels of compliance might have been in the absence of the rule change to determine what fraction of the compliance change, if any, should be attributed to the rule change. Examining data on compliance trends before a rule change help me approximate expected compliance levels after the rule change. Comparing these levels to levels of compliance with similar contemporaneous rules also helps me reduce the likelihood of falsely attributing causation to the rule. Analyses of counterfactuals build on such data and less quantitative assessments allow me to estimate the level of compliance without the rule change and gives me more confidence in the results. Taking these analytic steps together, allows me to identify significant variance in compliance across rules or after rules change and to identify when such variance can more plausibly be explained by the treaty and its rules and when it can more plausibly be explained by factors that were correlated with, but were independent of, the treaty.

Can lessons drawn from a study of rules within the single context of intentional oil pollution be generalized to other international environmental problems? The final chapter of this book addresses this question at length, but it deserves brief attention here. The issue of intentional oil pollution has many unique features, but also presents features common to other issues. It involves a pollution externality imposed on a global commons by a relatively concentrated industry in which actors are susceptible to regulation by both domestic and foreign authorities. Few environmental problems exhibit exactly these features, but many have one or more of them. For example, large power plants that emit sulfur dioxide and carbon dioxide that contribute to acid rain and climate change are similarly concentrated industries imposing pollution externalities on a global commons. Selecting a hard case involving opposition to many rules by some governments and industry as well as ongoing incentives for noncompliance makes lessons from these cases more likely to apply in cases in which these noncompliance pressures are weaker. More important, this book frames the lessons from the cases in terms that account for any influences unique to the oil pollution case. The study uses individual treaty provisions, rather than the treaty or issue itself, as the unit of analysis. Thus the ability to generalize from the findings here depends on which provisions are involved. For example, most environmental treaties have reporting provisions that are not dissimilar to those of the oil pollution treaties studied here.

There are also some factors that clearly limit the generalizability of the propositions developed here. This book draws attention to the role that dominant state concern played as a prerequisite for adoption of compliance-inducing oil pollution regulations. Although this concern is not sufficient to create a compliance system that works effectively, it provides one impetus to create an effective compliance information system, an impetus that is not always present. The nature of the regulated industry and the nature of the environmental problem itself impose constraints on the types of monitoring and enforcement processes available. Environmental problems created by less concentrated or internationalized industries, such as wetlands degradation, will provide different, and perhaps fewer, opportunities to alter the behavior of actors harming the environment. Oil pollution has been addressed strictly in a multilateral setting. However, bilateral environmental treaties undoubtedly contain mechanisms for specific reciprocity, for example, that have not been evident in a multilateral setting. Although the findings and propositions should apply to bilateral problems, the different dynamics of the latter may make many more opportunities available for resolution than those described here. Further discussion of these issues regarding how well the findings from intentional oil pollution can be applied to other environmental problems are provided in chapter 9.

To be sure, many strategies other than treaties can be adopted to make existing behaviors less environmentally malign. From ecosabotage, consumer boycotts, ecotourism, and debt-for-nature swaps to business-government contracts, education efforts, unilateral governmental initiatives, and efforts to reshape world civil society, advocates already have a range of potential remedies to international environmental problems.[45] We may hope that other creative new strategies will be developed. For the foreseeable future, however, treaties will be an essential and probably dominant tool in the international policy toolbox. This book provides one effort to better our understanding of the conditions necessary for their success so we can use them more effectively.

Outline of This Book

In chapter 2 I delineate and examine existing theories of treaty compliance, identifying the major factors hypothesized as explaining observed levels of international compliance. I highlight the shortcomings of current theory in addressing the two-level nature of enforcement in inducing compliance with environmental treaties. In chapter 3 I provide a historical background to the international treaties that regulate oil pollution. In that chapter I describe the nature of the intentional oil pollution problem, the basic interests and incentives of the government and industry actors involved, and the political background and context within which various changes have been made to improve compliance with oil pollution regulation.

In chapters 4, 5, and 6 I examine in detail how governments have responded to the various efforts to elicit improved compliance with provisions requiring reporting, enforcement, and the provision of oil waste reception facilities. In chapters 7 and 8 I report the results of similar analyses of the oil and shipping industries' responses to discharge limits and equipment standards aimed at inducing tanker operators to

45. For discussions of some of these strategies, see Elissa Blum, "Making Biodiversity Conservation Profitable: A Case Study of the Merck/INBio Agreement," *Environment* 35 (May 1993); Michael Martin, "Ecosabotage and Civil Disobedience," *Environmental Ethics* 12 (Winter 1990); Alan Patterson, "Debt for Nature Swaps and the Need for Alternatives," *Environment* 32 (December 1990); and Paul Wapner, "Making States Biodegradable: Ecological Activism and World Politics," unpublished manuscript, Washington, DC, 1993.

reduce intentional discharges of oil during tanker voyages. In chapter 9 I conclude the book by recapitulating the major empirical findings of my thesis, developing general recommendations regarding how treaties can be designed to elicit greater compliance even absent changes in the power and interests of the actors involved, and I delineate the contextual factors that condition the success of the recommendations and may limit their generalizability to other issues.

2
Compliance Theory: A Synthesis

Do nations and their citizens adjust their behavior to comply with environmental treaties? Can we improve environmental treaties to make compliance more likely? If so, how? Reliable answers to these and related questions require careful evaluation of past experience with efforts to improve such treaties. Such evaluation, in turn, profits from understanding the factors—treaty-based and otherwise—that previous scholarship suggests are sources of international behavior.

Policy interest in the relationship of treaties to behavior stems from a pragmatic concern that treaty goals are not always achieved as completely and effectively as possible. Government diplomats and lawyers spend considerable resources drafting and redrafting treaties to resolve international environmental problems. Environmental groups commonly support these efforts, pressing governments to negotiate new environmental treaties and to strengthen and refine existing treaties. Business groups regularly oppose provisions of environmental treaties as excessively costly and burdensome. Policy analysts and pundits regularly highlight the problems with existing treaties and propose new treaty provisions to address them. All these actions reflect a belief that better law can remedy bad behavior. They simultaneously illustrate an assumption that treaties can influence behavior and the fact that they do not always do so. Indeed, for many lawyers "the assumption that legal texts drive changes in behavior is second nature."[1] The number and variety of proposals to improve environmental treaties suggest, however, that we

1. Joan E. Donoghue, "On Learning Not to Think Like a Lawyer: Improving the Effectiveness of International Environmental Law," Washington, DC, unpublished manuscript, February 1993, 6.

still lack a solid understanding of what factors facilitate, and which impede, compliance with a treaty. Partly because the above-mentioned actors often become involved in the treaty process in response to concerns particular to a given treaty, they usually fail to identify sources of failure or success common to other treaties.

Political science research into the treaty/behavior relationship has received only limited attention historically, but is currently the focus of numerous studies as noted in chapter 1. In such international relations research, issues of compliance quickly enmesh one in a larger and long-standing debate over why nations behave the way they do. The realist school of thought, developed after World War II, views the pursuit and use of power and the anarchic structure of modern international relations as the primary determinants of international behavior. Realists consider international law as having little significant impact on nations' international policies. "Considerations of power rather than of law determine compliance" in all important cases.[2] Law influences behavior, if at all, only when relatively unimportant nonsecurity issues are at stake. The conformance of state behavior to treaty rules reflects spurious correlation rather than true causation: structural factors that lead states to take certain actions also lead them to negotiate treaties codifying those actions. Aspiring to explain "a small number of big and important things," realists have shown little interest in evaluating treaties as sources of international behavior.[3] Nevertheless, realism encourages a bias against assuming that treaties cause behavior to change, and provides an essential set of alternative explanations of why nations might take actions that conform to treaty provisions.

Institutionalists and international lawyers agree with Hans Morgenthau that "the great majority of the rules of international law are generally observed by all nations."[4] Disagreement arises over whether we can

---

2. Hans Joachim Morgenthau, *Politics Among Nations: The Struggle for Power and Peace,* brief ed., revised by Kenneth W. Thompson (New York: McGraw-Hill Inc., 1993), 268.

3. Kenneth Waltz, "Anarchic Orders and Balances of Power," in Robert Keohane, ed., *Neorealism and Its Critics* (New York: Columbia University Press, 1986), 328.

4. Morgenthau, *Politics Among Nations,* 267. See also, for example, Louis Henkin, *How Nations Behave: Law and Foreign Policy* (New York: Columbia

attribute such behavior to a treaty. For institutionalists, international institutions, regimes, organizations, and treaties "appear to be major determinants of collective behavior . . . at the international level."[5] Given this assessment, institutionalists have sought to identify the conditions under which treaties can influence behavior and the types of norms, principles, rules, and processes that do so most effectively.[6] While realists see states as dominating international affairs, nonstate actors also play important roles in institutionalist theories as targets of regulation and as participants in the effort to elicit compliance.[7] While institutionalists look for cross-treaty generalities regarding the causal links between behavior and treaties, they often fail to convert these into the practical advice demanded by policymakers.

Drawing on each of these three outlooks on compliance—pragmatist, realist, and institutionalist—in this chapter I develop a synthetic framework for subsequent analysis of the degree to which treaty rules influence behavior and the causal mechanisms by which they do so. Pragmatists demand that our framework be sufficiently well defined to allow for both empirical analysis of past experience and the development of prescriptive advice for future policy. Realists force us to rigorously question whether political or structural factors more readily explain, or at least condition, any correlations between treaty provisions and behaviors. Institutionalists identify the means by which treaties may influence behavior, demanding empirical validation of which treaties prove more or less effective. Together they provide propositions regarding the sources of compliance and noncompliance, the means by which treaty rules can increase compliance, and the exogenous factors that may also increase compliance.

University Press, 1979), 47; Oran Young, *International Cooperation: Building Regimes for Natural Resources and the Environment* (Ithaca: Cornell University Press, 1989), 62; and Abram Chayes and Antonia Chayes, "Compliance without Enforcement: State Behavior under Regulatory Treaties," *Negotiation Journal* 7 (July 1991), 31.

5. Young, *International Cooperation*, 61.

6. Stephen D. Krasner, ed., *International Regimes* (Ithaca: Cornell University Press, 1983).

7. For the seminal work in this area, see Robert O. Keohane and Joseph S. Nye, Jr., eds., *Transnational Relations and World Politics* (Cambridge, MA: Harvard University Press, 1972).

In this chapter I begin by defining compliance and noncompliance and delineating various possible sources of "first-order" compliance, that is, the reasons why national and subnational actors often comply even in the absence of any efforts to elicit compliance. In the chapter's second section I describe the opposing forces that may lead an actor to fail to comply. In the third section I outline how these pressures for noncompliance can be countered by the unilateral actions of governments and nongovernmental actors. In the last section of the chapter I describe how a treaty compliance system can be designed to elicit compliance. Finally I develop a framework for thinking about the various factors that institutionalists argue can improve compliance, identifying three components of a treaty compliance system—a primary rule system, a compliance information system, and a noncompliance response system—that form the basis for subsequent empirical analysis.

Definitions of Compliance and Noncompliance

I define *compliance* as an actor's behavior that conforms to a treaty's explicit rules.[8] As a subset of compliance, I distinguish treaty-induced compliance as behavior that conforms to such rules *because of* the treaty's compliance system. Using this definition, the realist-institutionalist debate becomes a question of whether treaty-induced compliance ever occurs. The term *compliance* is commonly applied in comparing behavior to specific treaty provisions, a treaty's broader spirit and principles, implicit international norms, informal agreements, and even tacit agreements.[9] Although ambiguous and nonexplicit rules, like principles and norms, may well influence behavior, empirically evaluating these influences usually founders on the inability to achieve agreement, often among the

8. Roger Fisher, *Improving Compliance with International Law* (Charlottesville, VA: University Press of Virginia, 1981), 20; and Oran Young, *Compliance and Public Authority: A Theory with International Applications* (Baltimore: Johns Hopkins University Press, 1979), 104.

9. See, for example, George W. Downs and David M. Rocke, *Tacit Bargaining, Arms Races, and Arms Control* (Ann Arbor, MI: University of Michigan Press, 1990); and Charles Lipson, "Why Are Some International Agreements Informal?" *International Organization* 45 (Autumn 1991). On ambiguity in treaty rules, see Abram Chayes and Antonia Chayes, "On Compliance," *International Organization* 47 (Spring 1993), 188–192.

parties and certainly among analysts, regarding whether a given action constitutes compliance or not. In contrast, restricting the study to *explicit* treaty provisions allows for replicable evaluation of compliance against clearer and less subjective standards. While recognizing that treaties may induce positive behavioral change that nonetheless fails to meet an established standard, I exclude from my definition the notion of compliance with the spirit of an agreement which, I believe, introduces an unnecessary element of subjectivity into any empirical analysis of compliance.

Evaluating compliance against treaty provisions also makes more sense than speaking of compliance with the treaty as a whole. Parties often comply with some treaty provisions while violating others. Within a nation, various actors—different government bureaucracies as well as industry and nongovernmental organizations—may well be responsible for implementing different treaty provisions.[10] To speak of "treaty compliance," therefore, causes one to lose valuable empirical information by aggregating violation of one provision with compliance with another. Compliance also needs to be kept distinct from the related concept of effectiveness. Although I discuss the distinction between the two at length in the final chapter, compliance can be thought of as exclusively a matter of altering behaviors without considering whether these behavioral changes are necessarily sufficient to accomplish the stated or unstated aims of the treaty. In most instances, compliance will correlate sufficiently with effectiveness to make more compliance preferable to less.

Sources of Compliance

Regulatory treaties' proscriptions of undesirable actions and prescriptions of desirable ones are often fulfilled. Determining whether the subsequent absence of proscribed actions or presence of prescribed ones is evidence of treaty-induced compliance would be simple if no other potential explanations for these behaviors existed. But, as noted above, the contention is not about whether nations comply, but why they comply. Governments and private actors face a wide variety of incentives and constraints in undertaking any action. Acts that the treaty defines as compliance (or violation) may be undertaken for numerous reasons

10. Chayes and Chayes, "Compliance without Enforcement," 318.

having little to do with treaty dictates. Enumerating the first-order sources of compliance and noncompliance, that is, their sources that are independent of treaty-related efforts to encourage compliance or deter violation, provides a valuable means of avoiding falsely characterizing compliance as treaty-induced.

For any given treaty or treaty provision, the actual compliance level across countries and across time is likely to reflect compliance due to a combination of sources. The compliance level will reflect the underlying structure of the environmental problem, the relationship of the treaty's requirements to existing behavior and future interests, and the structure and decision-making processes of the governments, corporations, and other organizations involved. A single environmental problem may pose different problems for developed countries and developing countries. In the case of many environmental treaty rules, some actors will unilaterally decide to comply, others will decide to violate, and others' decisions will depend upon whether other actors comply and how many comply. This section focuses on the factors that produce compliance even in the absence of a system to identify and respond to noncompliance, thereby highlighting both the exogenous forces for compliance and the important role that corporate as well as governmental actions play in international environmental treaties.

Compliance as Independent Self-interest

The easiest explanation of why a government or other actor regulated by a treaty undertakes a given behavior is because it believes the action furthers its interests. Nations often negotiate treaties precisely "for the promotion of their national interests, and to evade legal obligations that might be harmful to them."[11] As treaties are consensual agreements between nations, treaty provisions reflect the relative success of the different signatories in promoting their interests. Obviously a key determinant of a nation's willingness to comply is the degree of behavioral change the treaty requires. The degree of required change varies across treaties and across rules and actors for a single treaty.

Actors may comply because the treaty rules require no change in behavior. Through successful negotiation a country may place all the

---

11. Morgenthau, *Politics Among Nations*, 259.

burden of adjustment on other states. "Leader" states negotiating an environmental accord may already have established and implemented legislation that goes well beyond the requirements with which "laggard" states will agree. Industries already meeting a specified pollution standard may support treaties that require their foreign counterparts to do the same as a means of improving competitiveness without changing their own behavior. When agreements reflect policies that represent lowest common denominators, many states and companies will find themselves already in compliance. Some states may simply not be engaged, or be only minimally engaged, in the activity regulated by the treaty, as is the case with regard to Switzerland's participation in oil pollution and whaling activities.

Agreements sometimes proscribe undesirable actions that no actor currently has incentives to undertake with the hope of restraining future economic, political, or technological pressures to undertake such actions. Like the Antarctic Treaty, which imposes constraints on mining, such agreements codify existing behaviors to "protect against changes in preferences."[12] Compliance with these provisions of the Antarctic Treaty to date has been perfect because the availability of lower-cost sources elsewhere has meant that incentives to mine in Antarctica have remained low.[13]

States can also facilitate their own compliance by negotiating vague and ambiguous rules. Ambiguity may reflect agreements reached despite sincere differences about a specific rule's content—"papering over"—or efforts to accrue environmental praise by agreeing to terms that appear to require behavioral change, but actually prove sufficiently vague to allow for business as usual. The absence of an international court to authoritatively interpret such ambiguities naturally leads states to interpret treaty rules so they can behave as their interests dictate while claiming their behavior is in compliance.[14] Although excessively self-

12. Robert O. Keohane, *After Hegemony: Cooperation and Discord in the World Political Economy* (Princeton, NJ: Princeton University Press, 1984), 116.

13. M. J. Peterson, *Managing the Frozen South: The Creation and Evolution of the Antarctic Treaty System* (Berkeley: University of California Press, 1988).

14. Morgenthau, *Politics Among Nations,* 260; and Susan Strange, "Cave! Hic Dragones: A Critique of Regime Analysis," in Krasner, *International Regimes,* 350.

serving interpretations may well elicit criticism, ambiguous treaty language makes charges of outright violation difficult to prove.

When treaties require the adoption of new behaviors, they may require signatories to take only actions they already know they want to take. Unilateral compliance may be a preferred option. In some such cases, the state would behave as it does in any event, compliance being strictly coincidental. In other cases, the agreement provides international legitimacy, increasing domestic political support enough to enable the government to implement a desired but otherwise unattainable policy. For example, a climate change agreement may provide some governments with the impetus necessary to adopt energy taxes. Treaties may also reflect "suasion" games in which one or more powerful states or nongovernmental actors benefit from unilateral compliance but benefit more if others also comply.[15] Although it will seek to get others to comply, such a state or other actor will comply whether or not those strategies succeed. DuPont's phase-out of chlorofluorocarbons and Conoco's installation of double-hull tankers before it was internationally required suggest that these companies decided to comply independently of other companies' decisions, though they preferred that others comply.

The preceding sources of compliance show how actors may comply out of self-interest even if they define that interest myopically and independently of others' actions.[16] However, the calculus leading a state to comply may involve more expanded notions of independent self-interest than realist scholars would concede. Institutionalists point out that states sometimes adopt broader and longer-term views of self-interest, including joint gains and empathy, for example, that lead them to comply in a wider range of situations than realism would predict.[17] States and corporations may fear the unknown and unintended side effects of their current noncompliance on the future of the treaty and on a range of other

15. Lisa L. Martin, "Interests, Power, and Multilateralism," *International Organization* 46 (Autumn 1992); and Kenneth Waltz, *Theory of International Politics* (Reading, MA: Addison-Wesley Publishing Co., 1979), chapter 9.

16. Keohane, *After Hegemony*, 99.

17. Arthur A. Stein, "Coordination and Collaboration: Regimes in an Anarchic World," in Krasner, *International Regimes*, 138–139; and Keohane, *After Hegemony*, 120–124.

relationships.[18] They may fear adverse public opinion domestically or internationally. Parties may comply with rules viewed as fair and legitimate even if at times costly.[19] Even when hegemonic states coerce weaker states to accept a treaty, legitimate social purposes and changes in perceived self-interest may cause nations to continue complying past the point that immediate self-interest can explain.[20] These conceptions of self-interest veer away from strictly independent decision-making: rather than making worst-case assumptions that no others will comply, decision-makers use past experience to forecast compliance by others when calculating the benefits they expect from compliance.

Even if a treaty rule requires change, bureaucratic procedures, group think, and bounded rationality may make the choice of compliance, once initiated, hard to revisit. Governments and corporations deal with some compliance problems through standard operating procedures and habits, thereby foregoing potentially beneficial opportunities to violate in order to reduce overall decision-making costs, even in the case of issues involving national security.[21] Even realists admit that habit sometimes drives states to actions contrary to immediate self-interest because states do not constantly reassess their interests and power.[22] Businesses promulgate and train personnel in corporate procedures that reflect domestic and international laws, even in cases in which the likelihood that opportunistic violations will be detected is minuscule. International rules allow

18. See Keohane, *After Hegemony*, 105; and Donald Puchala and Raymond Hopkins, "International Regimes: Lessons from Inductive Analysis," in Krasner, *International Regimes*, 90.

19. Thomas M. Franck, *The Power of Legitimacy among Nations* (New York: Oxford University Press, 1990); and Puchala and Hopkins, "International Regimes," 66.

20. John Gerard Ruggie, "International Regimes, Transactions, and Change: Embedded Liberalism in the Postwar Economic Order," in Krasner, *International Regimes*, 200.

21. Abram Chayes develops a clear argument regarding how these factors lead to compliance in arms control agreements in his excellent article, "An Inquiry into the Workings of Arms Control Agreements," *Harvard Law Review* 85 (March 1972). See also Graham T. Allison, *Essence of Decision: Explaining the Cuban Missile Crisis* (Boston: Little, Brown, 1971); Young, *Compliance and Public Authority*, 25 and 178; and Young, *International Cooperation* 78–79.

22. Waltz, *Theory of International Politics*, 208, and Stein, "Coordination and Collaboration," 137.

actors to simplify or reduce the number of decisions they must make in a complex environment.[23]

Although compliance is calculated independently of other actors' behavior, it is not a static decision. Over time, economic and technological changes can "cause national governments to change their minds about which rules or norms of behavior should be reinforced and observed and which should be disregarded and changed."[24] Decreases in the price of alternatives to chlorofluorocarbons have increased the likelihood that the Montreal Protocol's phase-out deadlines will be met.[25] As will be discussed later, oil price shocks in the 1970s explain some of the increase in compliance with oil pollution treaty rules in effect at the time. In contrast, economic recession will reduce environmental compliance if countries redirect resources from environmental to development goals. Whether economic or technological shifts make compliance more or less attractive depends on the type of shift and government responses to it.[26]

Powerful nonstate actors, including multinational corporations, nongovernmental environmental groups, and scientists, often influence international politics both directly and by helping to define state interests.[27] New scientific knowledge or increased environmental activism can cause

23. Friedrich V. Kratochwil, *Rules, Norms, and Decisions: On the Conditions of Practical and Legal Reasoning in International Relations and Domestic Affairs* (Cambridge, England: Cambridge University Press, 1989), 14.

24. Strange, "Cave! Hic Dragones," 348; and Robert Gilpin, *War and Change in World Politics* (Cambridge, England: Cambridge University Press, 1981), chapter 2.

25. *Montreal Protocol on Substances That Deplete the Ozone Layer,* 16 September 1987, reprinted in 26 I.L.M. 1541 (1987), hereinafter cited as *Montreal Protocol.*

26. Realists, however, tend to assume that such changes work against compliance: "regimes are only too easily upset when either the balance of bargaining power or the perception of national interest (or both together) change among those states who negotiate them" (Strange, "Cave! Hic Dragones," 345).

27. Gilpin, *War and Change,* 51–52; Robert Gilpin, *U.S. Power and the Multinational Corporation: The Political Economy of Foreign Direct Investment* (New York: Basic Books, Inc., 1975); and Peter M. Haas, *Saving the Mediterranean: the Politics of International Environmental Cooperation* (New York: Columbia University Press, 1990).

increases in the perceived costs of an environmental externality and lead to greater levels of compliance. Elections or larger social or political factors often change the bargaining positions of domestic bureaucratic and political groups, altering how a state assesses its interests related to compliance. Domestic environmental groups may become increasingly powerful and concerned; treaties provide them with "a stronger case for constraint than would be possible in the absence of such obligations."[28] Such developments not only increase the domestic costs of violation, but often constrain even efforts at retaliatory noncompliance.[29] International agreements generate inertia that supports compliance once it has begun.[30] Such changes may increase overall compliance if they reflect transnational social shifts toward greater environmental concern.

At any given time, however, for those governments or nonstate actors that base compliance decisions on self-interests independently defined, compliance proves robust and concerns over noncompliance are minimal. Indeed, the behavior of these actors is not treaty-induced compliance. For these actors, treaty rules have been brought in line with existing or intended future behaviors, and not vice versa. When most parties to a treaty have such interests, the rule will be met with high levels of compliance even absent positive inducements or negative sanctions. Power plays little role in determining whether a state complies or not. Even without efforts to manipulate interests, we can expect considerable compliance simply because treaty rules reflect preexisting interests, the rules require little change in current behavior patterns, or the actors fail to recalculate their interests constantly. In these cases, compliance is not caused by the treaty, but merely coincides with it.

28. Jock A. Finlayson and Mark W. Zacher, "The GATT and the Regulation of Trade Barriers: Regime Dynamics and Functions," in Krasner, *International Regimes,* 312.

29. Kenneth Oye, "Explaining Cooperation under Anarchy: Hypotheses and Strategies," in Kenneth Oye, *Cooperation under Anarchy* (Princeton, NJ: Princeton University Press, 1986).

30. Stephen D. Krasner, *Structural Conflict: The Third World against Global Liberalism* (Berkeley: University of California Press, 1985), 29; James Rosenau, "Before Cooperation: Hegemons, Regimes and Habit-driven Behavior in World Politics," *International Organization* 4 (Autumn 1986); and Robert Jervis, "Realism, Game Theory, and Cooperation," *World Politics* 40 (April 1988).

Compliance as Interdependent Self-interest

Compliance can arise from interactive as well as independent decision-making.[31] States and corporations not only can include broader and longer-term concerns in their calculus of self-interest, but also can include their expectations regarding the impact their own compliance will have on others. Coordination and collaboration game models help clarify the operation of such interdependent conceptions of self-interest.

In coordination games, each actor prefers compliance so long as enough other actors comply. Although the meaning of *enough* varies from actor to actor, each assesses whether to comply based on the actions or expected actions of others. Realists see such complementarity of interests regarding compliance as explaining why most treaties require so little enforcement.[32] Like Schelling's "meeting" games, treaties can avert dilemmas of common aversion by coordinating action: the rule allows expectations to converge on an equilibrium behavior that, once states overcome the inertia that inhibits cooperation, none has incentives to violate.[33] The distribution of the benefits of compliance depends on the form of coordination but, once others choose to comply, the dominant strategy for all lies in complying.[34]

Coordination games do not face the "sanctioning problem" that plagues collaboration problems.[35] Actors are self-deterred because, while an actor's noncompliance hurts others, it hurts that actor enough to deter it. Besides, since noncompliance is always a public effort "to force the other actor into a different equilibrium outcome," detection problems do not arise.[36] Additionally, since other actors' most effective retaliation for

31. Stein, "Coordination and Collaboration," 117.

32. Morgenthau, *Politics Among Nations*, 267.

33. Thomas Schelling, *Micromotives and Macrobehavior* (New York: Norton, 1978).

34. Arthur A. Stein, *Why Nations Cooperate: Circumstance and Choice in International Relations* (Ithaca, NY: Cornell University Press, 1990).

35. Robert Axelrod and Robert O. Keohane, "Achieving Cooperation under Anarchy: Strategies and Institutions," in Oye, ed., *Cooperation under Anarchy*, 235.

36. Stein, "Coordination and Collaboration," 130.

noncompliance is continuing to comply, the desire to sanction reinforces rather than undermines the incentives to comply. For example, once nations agree on an allocation of the limited number of possible satellite orbits, no country has incentives to destroy its own satellite by placing it in the same orbit as another nation's and, if it were to do so, it could not do it clandestinely.[37] The distribution of power among nations determines *which* behavioral equilibrium nations adopt, with the strong dictating to the weak, but not whether the equilibrium will be maintained. Treaties will tend to codify the equilibriums that provide benefits to the strong, but they will not have the subsequent task of inducing compliance from weak states.

Situations involving public goods exhibit similar properties. If enough actors recognize that they can be better off collaborating to produce a public good and can trust each other enough to "jump" to this joint outcome, a subset of all actors can negotiate and comply with an agreement even though other nations continue to violate.[38] Actors willing to tolerate noncompliance by others may achieve joint gains, even though those gains are smaller than they would be if all actors complied. Actors that fail to consider how their decisions to comply encourage others to comply see the costs of compliance as outweighing the benefits. Actors that recognize that their decision to comply may convince enough others to comply to make the benefits outweigh the costs will attempt to assure others of their good faith. Common to both these scenarios is the fact that, given what one actor can dependably expect others to do, that actor's best strategy is to comply.

Unfortunately, environmental problems more often resemble collaboration games in which joint compliance is preferred to joint violation, but in which each actor's dominant strategy is to violate even if others comply. The literature on prisoners' dilemma, multiperson prisoners'

37. Stephen Krasner, "Global Communications and National Power: Life on the Pareto Frontier," *World Politics* 43 (April 1991).

38. Duncan Snidal, "The Limits of Hegemonic Stability Theory," *International Organization* 39 (Autumn 1985). See also Mancur Olson, *The Logic of Collective Action: Public Goods and the Theory of Groups* (Cambridge, MA: Harvard University Press, 1965), chapters 1 and 2; and Russell Hardin, *Collective Action* (Baltimore: Johns Hopkins University Press, 1982).

dilemma, and free riding is extensive.[39] Realists point out that the conditionality of benefits on compliance by others, coupled with the fear and uncertainty underlying international relations, makes collaboration essentially impossible. For example, as desirable as it may be collectively to reduce international industrial pollution, the anarchic international arena produces fears of free riding that lead states to define interests on unilateral, unconditional bases, thereby preventing agreement or compliance.[40] Despite the benefits of mutual compliance, the absence of international enforcement and risks of relative gains by others lead all parties to violate.[41]

Under such circumstances, compliance requires enforcement. International collaboration problems, in contrast to coordination problems, "must specify strict patterns of behavior and insure that no one cheats."[42] Unfortunately, if anything best characterizes the weakness of international law, it is the lack of an effective central enforcement system. Realists like Morgenthau note that "there can be no more primitive and no weaker system of law enforcement."[43] The institutionalists recognize that coordination games often face significant "sanctioning problems" in which the lack of ability and incentives to detect and respond to violation in turn hinder the achievement of compliance. Enforcement itself poses a collective action problem in which even states that comply may not monitor compliance or enforce rules against others.[44]

Compliance in the case of collaboration problems can arise from enforcement by a dominant or hegemonic state with system-wide concerns that sees what may be a collaboration problem for others as a suasion game. The dominant state manages the problem because it is

39. See, for example, Olson, *The Logic of Collective Action;* Hardin, *Collective Action;* and Robert Axelrod's *The Evolution of Cooperation* (New York: Basic Books, 1984).

40. Waltz, *Theory of International Politics,* 196–197.

41. Waltz, *Theory of International Politics,* 70.

42. Stein, "Coordination and Collaboration," 128–130.

43. Morgenthau, *Politics Among Nations,* 298.

44. Indeed since states that violate a treaty are unlikely to enforce it, unsuccessful enforcement efforts only increase the relative gains of noncompliant states. As with compliance, models of limited rationality or habit-driven behavior may best explain the variance between those that enforce and those that do not.

capable of, and perceives sufficient benefits from, complying itself and/or enforcing compliance by others. Weak states are forced to comply with these "imposed orders" by "coercion, cooptation, and the manipulation of incentives."[45] The economic and technological changes mentioned earlier can alter compliance levels by increasing a dominant state's power to enforce and interest in enforcing such treaties. Growth in international economic interdependence will increase a dominant state's desire and capacity to exercise control of the international system.[46] Improved satellite surveillance could aid a dominant state's ability to monitor activities and impose sanctions more swiftly, thereby increasing compliance. Such changes may correlate with treaty amendments if they also lead to compliance system changes, such as the use of less ambiguous wording. In such cases, however, tighter wording is simply another indicator of the changed interests that have caused improved compliance rather than being an independent cause of improved compliance.

Fears of free riding can also be overcome if states view the benefits they will derive in other existing and future international agreements as conditional upon a record of compliance.[47] Such caution is fostered when states detect violations and either reciprocate with their own violations or "discount the value of agreements on the basis of past compliance."[48] Even if compliance in a given instance may be costly when narrowly construed, the costs in other areas and to its reputation can induce a state to comply. Compliance under such conditions is possible, but will be more fragile and more difficult to establish than when actors have independent interests in compliance. Although some actors may comply, efforts to elicit compliance by reluctant actors may also be required.

Since such cooperation depends on some degree of international trust, decreases in underlying rivalries and competition, such as those seen at

---

45. Oran Young, "Regime Dynamics: The Rise and Fall of International Regimes," in Krasner, ed., *International Regimes*, 100.

46. Waltz, *Theory of International Politics*, 209.

47. Keohane, *After Hegemony*, 105–106; Axelrod and Keohane, "Achieving Cooperation Under Anarchy," 250; and Young, *International Cooperation*, 75–76.

48. Keohane, *After Hegemony*, 105; and Morgenthau, *Politics Among Nations*, 267.

the end of the Cold War, will likely increase compliance levels.[49] As with changes in power, such a change may produce changes in treaty rules as well as compliance levels, leading to spurious correlations. However, the regime itself can increase trust by improving knowledge and reducing misperceptions of other states.[50] Over time trust, reputation, rule legitimacy, and habitual practice grow and can reinforce incentives for compliance.

Sources of Noncompliance

Having noted several reasons for compliance, we can now turn to the reasons for noncompliance. Especially in collaboration situations, absent efforts by other parties to encourage their compliance or discourage their violation, some actors may find it preferable to violate a given treaty rule. Other factors may also lead to noncompliance. To understand how actors elicit compliance in the face of such preferences, one must be aware of the various sources of noncompliance.

Noncompliance as a Preference

An actor may prefer noncompliance simply because the benefits of compliance, absent coercive efforts, simply do not outweigh its costs. This situation may arise for several reasons. Some actors may consciously sign treaties to garner the political benefits of membership, never intending to comply.[51] Others may feel strong domestic and international pressures to sign an agreement regardless of the costs of compliance. For example, the 178 countries in attendance at the 1992 United Nations Conference on Environment and Development faced strong pressures to sign climate change and biodiversity agreements; few countries refused to sign based on claims that compliance did not serve their interests. Other states may view most, but not all, rules in a treaty as in their interests, leading them to sign with the intention of complying with most, but not all, rules.

49. Stein, *Why Nations Cooperate*, 187; and Axelrod and Keohane, "Achieving Cooperation under Anarchy," 227.

50. Axelrod and Keohane, "Achieving Cooperation Under Anarchy," 247.

51. It remains an empirical question whether this is common or "exceptional" international behavior (Chayes and Chayes, "Compliance without Enforcement," 311).

Assumptions that states will comply because they have signed ignore the fact that a state's material interests may include signature but not compliance. At least three scenarios can account for this. First, a state may be a classic free rider, valuing the benefits of compliance by others, but seeking to avoid the costs of its own compliance. Second, a state may value compliance by itself and by others and even deem that the benefits of compliance outweigh the costs, but may nonetheless prefer to devote the compliance-related resources to more pressing social problems. Third, a state may view compliance as having no real benefits. For instance, a state may not value actions that it admits would improve the environment. It is tempting to view environmental problems as global commons or prisoners' dilemma problems with the implicit assumptions that all actors prefer mutual compliance to mutual violation. Yet this view ignores basic disagreements over the relationship of humans to the environment. Iraq's intentional dumping of oil during the Gulf War, the controversy over commercial whaling, and the general debate over environment versus development reflect such disagreements.

Even when total social costs make compliance a preferred strategy, the incidence of those costs usually means that those being asked to comply will have continuing incentives to violate. Environmental treaties frequently seek to remedy industrial pollution externalities; by definition, therefore, compliance involves companies' incurring costs that they had previously imposed on other social groups. Compliance with the Long-Range Transboundary Air Pollution Convention's restrictions on sulfur dioxide emissions required private power plants to install scrubbers and otherwise increase costs so citizens in other countries could enjoy a cleaner environment.[52] The resources that targeted industries use to oppose international regulatory efforts attest to the costs they view as involved in compliance. Whatever actors are involved, whether reluctant nations or reluctant industries within a nation, benefits from compliance are realized only when others link compliance to issues other than

---

52. *Convention on Long-Range Transboundary Air Pollution,* 13 November 1979, reprinted in 18 I.L.M. 1442 (1979), hereinafter cited as *LRTAP Convention;* and *Protocol to the 1979 Convention on Long Range Transboundary Air Pollution on the Reduction of Sulphur Emissions or Their Transboundary Fluxes by at Least 30 Percent,* 9 July 1985, UN Doc. ECE/EB.AIR/12, reprinted in 27 I.L.M. 707 (1988), hereinafter cited as *LRTAP Sulphur Protocol.*

immediate environmental concerns of the treaty through positive or negative inducements.

Noncompliance Due to Incapacity

Even actors that perceive compliance as beneficial may fail to comply for lack of the necessary resources. A willingness to pay does not necessarily equate to an ability to pay. Violation can be due to financial, administrative, or technological incapacities rather than any unwillingness to comply.[53]

While the governments of developing countries may not value environmental improvement because of a lack of domestic constituencies or because of more pressing concerns, even those that do may simply not have sufficient resources to meet the cost of compliance. Growing awareness of this source of noncompliance has led nations to establish mechanisms to finance compliance, as in both the London Amendments to the Montreal Protocol and the Framework Convention on Climate Change.[54]

Environmental noncompliance also results from a lack of administrative capacity.[55] Even if we assume that governmental decisions to comply with treaty rules will bring government actions into compliance, that assumption comes into serious question when compliance requires a government to successfully alter the actions of myriad subnational actors. This two-level quality of environmental treaty compliance can lead to noncompliance when a government lacks informational or regulatory infrastructures adequate to elicit compliance. Even using its best efforts, the Brazilian government may fail to successfully communicate restrictions on tree clearing to the peasant farmers responsible, leading to

53. Oran Young, "The Effectiveness of International Institutions: Hard Cases and Critical Variables," in James N. Rosenau and Otto Czempiel, eds., *Governance without Government: Change and Order in World Politics* (New York: Cambridge University Press, 1991), 183–185.

54. Article 10 of *London Amendments to the Montreal Protocol on Substances that Deplete the Ozone Layer,* 29 June 1990, reprinted in 30 I.L.M. 539 (1991), hereinafter cited as *London Amendments;* and Article 4(3) and Article 11, *Framework Convention on Climate Change,* UN Document A/AC.237/18 (part II)/add.1 (15 May 1992).

55. See the introduction and conclusion of Peter Haas, Robert O. Keohane, and Marc Levy, *Institutions for the Earth: Sources of Effective International Environmental Protection* (Cambridge, MA: The MIT Press, 1993).

noncompliance. Even if the informational infrastructure exists, an effective regulatory infrastructure that can induce behavioral change by these actors may be absent. Tankers registered in Liberia and Panama rarely enter those countries' ports, making inspections to monitor compliance with international shipping standards difficult.

Negotiators may even establish standards that current technologies are unable to meet. Hopes that regulatory necessities will lead to technological inventions do not always prove well founded, leaving companies with no means of complying or only prohibitively expensive means. While these problems often boil down to variants of financial incapacity problems, cultural and social contexts may make compliance significantly more difficult to elicit from the companies and citizens of one country than from those of another.[56]

Noncompliance Due to Inadvertence

Finally, states may take actions with the sincere intention and expectation of achieving compliance, but nonetheless fail to meet treaty standards. Environmental rules establishing aggregate national targets for pollution reduction by specified deadlines may pose particular problems in this regard.[57] For example, a carbon tax established in good faith at a level deemed sufficient to achieve a 20 percent reduction in carbon dioxide emissions by a specified date might reduce emissions by only 15 percent by that deadline.[58] This problem is not restricted to developing states; the inherent uncertainty of the impacts of most policy strategies make it possible that even developed states' efforts to alter their citizens' and companies' behaviors will fail to achieve their intended results.[59] And

56. For example, two governments could be equally committed to reducing population growth, but one with a predominantly Catholic population might prove significantly less successful in complying.

57. Examples include the *Montreal Protocol* and the *LRTAP Convention*.

58. For a discussion of the inherent uncertainties surrounding actual emission reductions for a given tax level, see Joshua M. Epstein and Raj Gupta, *Controlling the Greenhouse Effect: Five Global Regimes Compared* (Washington, DC: The Brookings Institution, 1990), 15–17.

59. A seminal study on these issues is Jeffrey L. Pressman and Aaron Wildavsky, *Implementation: How Great Expectations in Washington Are Dashed in Oakland, or, Why It's Amazing that Federal Programs Work at All, This Being a Saga of the Economic Development Administration as Told by Two Sympathetic*

programs that bring one country into compliance may fail to have the same results elsewhere. Whether due to misguided policy or to the inherent uncertainties in outcomes of certain policies, the two-level nature of environmental compliance may make compliance due to inadvertence especially common.

Eliciting Compliance in the Face of Pressures for Noncompliance

To this point, our discussion has largely been restricted to first-order sources of compliance and noncompliance, that is, those incentives that governments and private actors have to comply with treaty rules before efforts are made to directly influence their choice. Whether these first-order incentives actually lead a government or a company to comply or violate depends on how willing and able other actors are to manipulate incentives to make compliance both possible and preferable. Such manipulation can involve positive inducements or negative sanctions that take the treaty framework as a given, or they can involve refining various components of the treaty compliance system itself.

Inducements

Giving positive rewards for compliance provides one means of increasing incentives for compliance. To the extent that noncompliance is due to inadvertence or incapacity, other actors can respond to those that appear likely to violate or already have violated a treaty rule in two ways. Education can clarify treaty requirements and identify strategies for compliance. Educational efforts can involve diplomatic discussions between government officials or can directly target the private actors responsible for the environmental problem. Such efforts help avoid noncompliance due to lack of knowledge or inadvertence. Industry groups in one country often conduct seminars to educate their own personnel as well as those of companies in other countries as new regulations are promulgated. Such seminars can demonstrate and disseminate information on cheap alternative means of complying that help

---

Observers Who Seek to Build Morals on a Foundation of Ruined Hopes (Berkeley: University of California Press, 1973).

increase the ranks of compliers. Private corporations that wish to increase sales may seek out and promote compliant technologies, independent of any governmental or intergovernmental efforts. As environmental concerns increasingly influence overseas development aid, policy advisers are helping devise programs to address or avoid the administrative incapacity problems that plague many developing nations.

Financial transfers can elicit greater levels of compliance when noncompliance arises due to any of the factors delineated above. It essentially entails one actor's paying for another's compliance. For instance, Finland paid the former Soviet Union to clean up a nickel smelting plant.[60] Intergovernmental side-payments outside of formal treaty procedures for environmental treaty compliance have been infrequent to date. Nongovernmental organizations (NGOs) have paid debts for developing countries and pharmaceutical companies have negotiated deals with them to create environmental preserves.

The fact that these actions have not been linked with specific environmental treaty provisions suggests two problems. First, governments prove reluctant to pay not only their own compliance costs, but those of other governments that are obligated under the treaty to comply in any event. Second, such efforts, whether taken within or outside the treaty context, face the "mundane problem of funding."[61] Financing proves difficult to organize because it poses collective action problems for those providing funds.[62]

Sanctions

The more traditional remedy for noncompliance has involved deterrence through the threat or use of sanctions. To be effective these sanctions

---

60. See Timothy B. Hamlin, "Debt-for-nature Swaps: A New Strategy for Protecting Environmental Interests in Developing Nations," *Ecology Law Quarterly* 16 (1989); Elissa Blum, "Making Biodiversity Conservation Profitable: A Case Study of the Merck/INBio Agreement," *Environment,* 35 (May 1993); and Anonymous, "Soviet Union Gives Finnish Metal Company Go-ahead to Modernize Two Nickel Smelters," *International Environment Reporter,* 10 October 1990, 421.

61. Chayes and Chayes, "Compliance without Enforcement," 318.

62. Keohane, *After Hegemony,* 159.

must be both credible and potent. Actors hoping to encourage compliance must convince reluctant actors that the likelihood that a violation will be detected and stiffly sanctioned makes the expected costs of violation exceed those of compliance.

Governments, corporations, NGOs, and general publics can all impose sanctions. Actors targeted for sanctions can similarly include governments, corporations, or even individuals.[63] Social opprobrium and world public opinion can pressure such actors to comply through several channels.[64] Diplomats are "called to account by other participants in the international system, either through bilateral diplomacy or in international forums."[65] NGOs and the publics they mobilize can prompt letter-writing campaigns, consumer boycotts of imports and tourism, corporate cancellation of contracts, and government trade restrictions, as evidenced by the response to Norway's resumption of commercial whaling.[66] Whether these actions are adequate to induce a reluctant business or government to comply, however, is open to interpretation.

Realists contend that public opinion exerts no "restraining influence upon the foreign policies of national governments"; institutionalists hold that it "can have unexpectedly powerful results . . . help[ing] to bring about far-reaching [human rights] changes in powerful and entrenched regimes."[67] The former see collective goods as supplied when "the interest of preeminent powers in the consumption of collective goods is strong

63. For example, individual traders may be convicted and punished for trades is endangered species prohibited by the *Convention on International Trade in Endangered Species of Wild Fauna and Flora*, 3 March 1973, T.I.A.S. no. 8249, 27 U.S.T. 1087, 983 U.N.T.S. 243, reprinted in 12 I.L.M. 1085 (1973), hereinafter cited as *CITES*. For an analysis of the CITES agreement, see Mark C. Trexler, "The Convention on International Trade in Endangered Species of Wild Fauna and Flora: Political or Conservation Success?" unpublished Ph.D. thesis, University of California-Berkeley, Berkeley, CA, December 1989.

64. Young, *Compliance and Public Authority,* 22 and 44.

65. Abram Chayes and Antonia Chayes, "Adjustment and Compliance Processes in International Regulatory Regimes," in Jessica Tuchman Mathews, ed., *Preserving the Global Environment: The Challenge of Shared Leadership* (New York: W. W. Norton and Co., 1991), 290.

66. Michael D. Lemonick, "The Hunt, the Furor," *Time* 2 August 1993.

67. Morgenthau, *Politics Among Nations,* 265; and Chayes and Chayes, "Compliance without Enforcement," 324.

enough to cause them to undertake the provision of those goods without being properly paid."[68]

A dominant state must both establish and enforce treaties resolving collective action problems.[69] Depending on its own commitment to the issue, which may vary over time, the dominant state may choose to tolerate violations, expend resources to force others to comply, or expend resources to force others to both comply with and enforce the treaty. In such cases, weak states comply to the extent that stronger states force them to—or credibly threaten to force them to—via political and economic sanctions for violation.[70] Compliance does not require explicit enforcement: the prestige or reputation for power of stronger states leads weaker states to comply with rules that conflict with short-term self-interest even absent explicit enforcement.[71] Powerful states, and they alone, use sanctions to enforce those international rules that suit their immediate interests.[72] In essence, states do not enforce international law, but command obedience from weaker states.[73]

Compliance under such a model of hegemonic enforcement will tend to decline as powerful states lose their power monopoly and other states with different interests gain power.[74] Also, the costs of maintaining international agreements tend "to rise faster than the financial capacity

68. See Waltz, *Theory of International Politics,* 208; Gilpin, *War and Change,* 169; and Mancur Olson, *The Logic of Collective Action* (Cambridge, MA: Harvard University Press, 1965), 35.

69. See Gilpin, *War and Change,* 30; Waltz, *Theory of International Politics,* 136 and 157; and Krasner, "Structural Causes and Regime Consequences: Regimes as Intervening Variables," 13–15.

70. Gilpin, *War and Change,* 36. Strange suggests that compliance with multilateral treaty rules, e.g., GATT, is best explained by parallel bilateral deals that allow for the more efficacious and direct application of power (Strange, "Cave! Hic Dragones," 352).

71. Gilpin, *War and Change,* 31.

72. Morgenthau argues that attempts by weak states to violate, by strong states to enforce, and the success of both, depend on "political considerations and the actual distribution of power" (*Politics Among Nations,* 266).

73. Young defines "obedience as a response to a command from an identifiable authority figure in a specific situation" (*Compliance and Public Authority,* 5).

74. Waltz, *Theory of International Politics,* 177–178. See also Gilpin, *War and Change,* 173–174.

of the dominant power to support its position."[75] Thus enforcement demands become greater at the same time that the power to enforce lessens.[76] Since states violate treaty commitments whenever they have the interest and power to do so, decreased hegemony increases noncompliance as weak states gain relative power.

An alternative view of hegemonic power's influence on compliance levels suggests that, as extreme power asymmetries decrease, compliance increases because no actor possesses sufficient power "to ignore the dictates of the resultant institutions with impunity."[77] Likewise, weaker states can no longer expect a public good to be provided without their contribution, and therefore they comply. Although they predict different effects of reduced asymmetries of power on compliance, both theories suggest that changes in power distribution will produce changes in compliance.

Reciprocity has been the dominant factor institutionalists have used to explain the "potential anomaly" of why governments comply "with rules that conflict with their myopic self-interest."[78] In collaboration problems, each party has immediate and continuing incentives to violate and few incentives to enforce the rules against others. Rather than general concerns over reputation, discussed above, specific reciprocity—promising to comply if others comply and threatening to violate if others violate—provides an enforcement strategy. If actors have long time horizons, regular and continued interactions, and rapid and reliable information about the actions of others, reciprocity strategies can overcome the problem of the noncompliant free rider.[79] For example, reciproc-

75. Gilpin, *War and Change,* 156.

76. Charles Lipson, "The Transformation of Trade: The Sources and Effects of Regime Change," in Krasner, ed., *International Regimes,* 254–255.

77. Young, "The Effectiveness of International Institutions," 187.

78. Keohane, *After Hegemony,* 99. For extended discussions of reciprocity, see Robert O. Keohane, "Reciprocity in International Relations," *International Organization* 40 (Winter 1986); and Axelrod, *The Evolution of Cooperation.*

79. Axelrod and Keohane, "Achieving Cooperation under Anarchy," 232. See also Axelrod, *The Evolution of Cooperation;* Robert O. Keohane, "The Demand for International Regimes," in Krasner, ed., *International Regimes;* and Robert Jervis, "Security Regimes," in Krasner, ed., *International Regimes.*

ity strategies have been used to explain compliance with arms control treaties.[80]

Although it works well bilaterally, reciprocity "may not prove compelling in a multilateral situation."[81] The initial violation may not impose sufficient costs on any single actor to provide incentives for retaliation. At the same time, retaliatory violations that do occur impose costs on compliers and violators alike, leading the retaliator to "suffer the opprobrium" of compliant actors while failing to impose sufficient costs on the initial violator to induce compliance.[82] These disincentives to retaliation reduce the credibility essential to the strategy and can lead to escalatory violations that undermine compliance by all parties.[83]

Actors can skirt the problems inherent in retaliatory noncompliance through sanctions that link environmental compliance to other issues. In some instances the gains of mutual compliance are sufficiently large that some actors undertake enforcement. Governments can use a wide range of unilateral or coordinated efforts at trade or other economic sanctions. Such sanctions, whether undertaken by states or other actors, allow costs to be targeted at the initial noncomplier, avoiding the risks outlined above of retaliatory noncompliance. NGOs and publics often prove more willing than governments to sanction violations by corporations or other governments. Governments can avoid retaliation by targeting sanctions against violating nationals or corporations of other countries; governments that might defend their own wrongful actions often care less about defending those of their nationals. Thus governments can induce compliance by foreign nationals without harming intergovernmental relationships regarding other issues by pressuring the other government directly.

In short, even if the treaty context is taken as a given, many different actors can take a wide array of actions to induce actors to comply despite their predisposition to violate.

---

80. Young, *Compliance and Public Authority*, 58; and Donald Wittman, "Arms Control Verification and Other Games Involving Imperfect Detection," *American Political Science Review* 83 (September 1989).

81. Keohane, "Reciprocity," 12.

82. Keohane, "Reciprocity," 12.

83. John Conybeare, "Trade Wars: A Comparative Study of Anglo-Hanse, Franco-Italian, and Hawley-Smoot Conflicts," in Oye, *Cooperation under Anarchy*, 151.

Compliance System Design[84]

Given the realist skepticism regarding the impact of treaties on behavior, we must rely on institutionalists and pragmatists to identify principles for designing a treaty compliance system that will elicit high levels of compliance. The institutionalists' disagreement with realists is not that other factors do not cause changes in compliance. Rather they make the narrower claim that treaty rule changes can also increase compliance. While regimes and treaty rules reflect power and interests, they also have an independent effect on compliance. Power and interest sometimes provide inadequate explanations of observed correlations between international rules and actors' behavior.[85] International behavior can reflect not only compliance, but treaty-induced compliance. However, whether a particular treaty actually induces compliance depends on how its compliance system is designed. Critical to this argument is the notion that the structure of international power and interests underdetermines a treaty's compliance system; it assumes that at a given point in time nations could have agreed to more than one compliance system design.

A compliance system is that subset of the treaty's rules and procedures that influence the level of compliance with a given rule.[86] Building on this concept, I distinguish three subsystems of any compliance system: a primary rule system, a compliance information system, and a noncompliance response system. If variance in compliance exists across treaties that cannot be explained by factors exogenous to the regimes, these three systems provide a framework for identifying the source of such variance in inducing compliance.

84. For a more extended discussion of compliance system design, see Ronald B. Mitchell and Abram Chayes, "Improving Compliance with the Climate Change Treaty," in Henry Lee, ed., *Controlling Climate Change: The Political Economy of Mitigation Strategies* (Washington, DC: Island Press, forthcoming).

85. Young, however, cautions against attributing too much power to institutions (*Compliance and Public Authority,* 44), as do Jock Finlayson and Mark Zacher ("The GATT and the Regulation of Trade Barriers," in Stephen D. Krasner, ed., *International Regimes,* 311–312), and Puchala and Hopkins ("International Regimes," 86–87).

86. Young, *Compliance and Public Authority,* 3.

The primary rule system consists of the actors, rules, and processes related to the behavior that is the substantive target of the regime. By its choice of who is regulated and how, the primary rule system determines the degree and sources of pressures and incentives for compliance and violation. The compliance information system consists of the actors, rules, and processes that collect, analyze, and disseminate information regarding violations and compliance and the parties responsible. The self-reporting, independent monitoring, data analysis, and publishing activities that comprise the compliance information system determine the amount, quality, and use of data on compliance and enforcement. The noncompliance response system consists of the actors, rules, and processes that govern the formal and informal responses undertaken to induce those identified as in noncompliance to comply. The noncompliance response system determines the type, likelihood, magnitude, and appropriateness of responses to noncompliance.

An environmental treaty's primary rule system, in contrast to that of an arms control or human rights treaty, generally attempts to alter the actions of private, subnational actors through the implementation activities of national governments. Environmentally activist governments must elicit behavioral change by the primary target of regulation—for example, oil and shipping companies—either directly or by prompting the responsible national governments to implement and enforce treaty terms. Frequently the rules and procedures comprising all three components of the compliance system can address governmental, nongovernmental, or subnational entities.

What factors endogenous to the treaty could increase compliance? What changes in the treaty compliance system would make compliance more likely for a given set of actors when addressing a given environmental problem? What attributes of treaty rules and processes should we look for as the basis of its impact? Compliance system design involves two distinct elements that address these questions. First, the selection of the primary rules determines how many actors are predisposed toward compliance and how many toward noncompliance. Whether it is made consciously or not, the choice of how the treaty is to define compliance and which actors must change their behavior becomes a major determinant of compliance. It sets the bounds on how many actors will comply voluntarily and the efforts needed to get others to comply. Second, having selected certain primary rules, the system must maximize the likelihood that actors will respond to noncompliance through means that redress the source of noncompliance. This element of design involves attempting to achieve the highest level of compliance possible by improving the compliance information and noncompliance response systems to make the detection of noncompliance and the response to it more likely, credible, and potent.

Primary Rule System

The preceding discussion has highlighted how the incentives of actors to comply with or violate the provisions of a treaty depend critically on the structure of the problem the treaty is meant to solve. Collaboration problems fascinate analysts precisely because their resolution requires jointly desirable compliance that proves hard to generate internationally.[87] Different approaches to the resolution of such problems greatly influence who bears what costs under what conditions and the degree to which benefits depend on actions by other actors. Given this fact, "choices of strategies and variations in institutions are particularly important, and the scope for the exercise of intelligence is considerable."[88]

Although some analysts have explored how institutions can alter payoff structures, fewer have explored how nations can choose between solutions to the same problem that have different payoff structures. Payoff structures may vary by solution as well as by problem. Even if international institutions prove too weak to alter payoff structures, they may be able to increase compliance by selecting from an array of possible solutions those that increase the likelihood of compliance.[89] For a given problem, several alternative solutions may exist. Even if, under conditions of perfect compliance, these solutions provided equal environmental benefits, they might exhibit different likelihoods of compliance because they regulated different actors. What kind of activity is being regulated? How many actors with what interests must change their behavior? How

87. Stein, "Coordination and Collaboration."

88. Axelrod and Keohane, "Achieving Cooperation under Anarchy," 228–232.

89. As a student of Jervis noted, "what . . . are we to cooperate *about*?" See Robert Jervis, "Realism, Game Theory, and Cooperation," *World Politics* 40 (April 1988), 331.

large and costly a change is involved? What exogenous incentives to comply do these actors have? By answering these questions through the process of selecting the point at which there should be regulatory intervention and defining the standards for compliance, negotiators may well have already largely determined the degree of treaty compliance. Claims that inducing compliance with the Framework Convention on Climate Change will be harder than inducing compliance with the Montreal Protocol rest on the notion that more people and industries must make much bigger behavioral changes to comply with the former. Regional agreements seek to increase cooperation by increasing the frequency of interaction between actors. Agreements that put greater burdens on developed countries than on undeveloped countries may generate more compliance because those countries have the incentives and resources to comply. In short, the selection of primary rules may prove the most powerful lever international policymakers have over the level of compliance elicited.

One key feature of the selection of a regulated activity involves its transparency. Some activities are highly visible and involve transactions between actors, while others are autonomous acts performed largely out of view. While transparency's main contribution to compliance is through making detection of violations easier, as discussed below, it also facilitates compliance by reassuring each actor about others' behavior. Actors predisposed to comply but concerned about noncompliance by others will violate under conditions of uncertainty. The ranks of "nice" actors—actors that will initially comply with a treaty rule before having information about the actions of others[90]—can be swelled by reducing this uncertainty. By regulating more transparent actions, treaties assure actors that others' noncompliance will be immediately visible and thus permit them to protect their interests accordingly. Reducing such fears that produce initial noncompliance allows compliance to develop.

Besides making strategic choices of who and what to regulate, primary rules can increase compliance through greater specificity.[91] Chayes urges

90. Axelrod, *The Evolution of Cooperation.*

91. Fisher, *Improving Compliance with International Law;* Duncan Kennedy, "Form and Substance in Private Law Adjudication," *Harvard Law Review* 89 (1976); and Joan E. Donoghue, "On Learning Not to Think Like a Lawyer."

"more concrete and quantitative performance criteria as they become politically and empirically validated."[92] Increasing specificity increases compliance in at least two ways. First, for actors disposed to comply, specific rules make compliance easier by reducing the uncertainty about what they need to do to comply.[93] Specific rules also reassure actors that others will not dispute the compliance of a given act. The actor can therefore act without fear of facing sanctions for noncompliance despite a good-faith effort to comply. Second, for actors predisposed not to comply, precise treaty language removes the excuse of inadvertence and misinterpretation when such actors must account for noncompliance.[94] Greater specificity will not change the actions of any actor facing significant compliance costs, but it may alter the choices of actors at the margin of compliance.

The notion that greater specificity improves compliance assumes that specificity can be increased without any change in the parties' willingness to agree among themselves. At any given point in time, it may prove impossible for actors to agree on more precise language. However, a rule's level of precision need not always correspond with the maximum level of precision that would reflect the existing level of common interest.

92. Abram Chayes, "Managing the Transition to a Global Warming Regime or What to Do Till the Treaty Comes," *Foreign Policy* 82 (Spring 1991), 16.

93. There is considerable debate over whether formal treaties provide more or less likelihood of behavioral change. Puchala and Hopkins conclude that the degree of formalization of a regime and its rules "have relatively little to do with the . . . probabilities of participants' compliance" ("International Regimes," 88). Charles Lipson has recently written an excellent article on the value of informal agreements ("Why Are Some International Agreements Informal?"). In contrast, Abram Chayes and Antonia Chayes argue that "it seems clear that an international regulatory regime will stand a better chance of achieving its objectives if it's based on a formal treaty"; see "Non-proliferation Regimes in the Aftermath of the Gulf War," in Joseph Nye and Roger Smith, eds., *After the Storm: Lessons from the Gulf War* (Lanham, MD: Madison Books and the Aspen Strategy Group of the Aspen Institute, 1992), 50.

94. Louis Kaplow, "Optimal Deterrence, Uninformed Individuals, and Acquiring Information About Whether Acts Are Subject to Sanctions," *Journal of Law, Economics, and Organization* 6 (Spring 1990) 93–128; and Louis Kaplow, "The Optimal Probability and Magnitude of Fines for Acts That Definitely Are Undesirable," *International Review of Law and Economics* 12 (March 1992), 3–11.

Even within a context of nations' seeking to avoid major constraints on their behavior, the range of negotiable outcomes may include some rules that are less vague than others, making the choices between rules important to subsequent compliance levels.

Compliance Information System

The major goal of any treaty's compliance information system is to maximize transparency. *Transparency* refers to both the amount and quality of the information collected on compliance or noncompliance by the regulated actors as well as the degree of analysis and dissemination of this information. Increasing transparency is seen as an essential component of any prescription to increase compliance.[95] Transparency is essential to hegemonic states enforcing compliance and to states resolving coordination problems, and it provides the reciprocity needed to elicit compliance from states when resolving collaboration problems.[96] To make the threat of a retaliatory violation—or linkage of compliance to other issues via sanctions or inducements—credible, the regulated actors must know that their choices will not go unnoticed.

If power and interests leave more than one rule in the zone of possible agreement, then negotiators' choices between rules will affect how easy verification is.[97] The choice of primary rules plays a crucial role since the choice of what acts to regulate implicitly includes a choice of how transparent compliance and noncompliance will be. The Montreal Protocol regulated the production of chlorofluorocarbons because it was far easier to monitor a very few producers than thousands of consumers. Although numerous activities contribute to climate change, technological capabilities will make verification of point-source polluters, like power

95. See, for example, Young, "The Effectiveness of International Institutions"; and Chayes and Chayes, "Adjustment and Compliance Processes."

96. Robert Jervis, "From Balance to Concert: A Study of International Security Cooperation," in Oye, ed., *Cooperation under Anarchy,* 74. See also George W. Downs, David M. Rocke, and Randolph M. Siverson, "Arms Races and Cooperation," in Oye, ed., *Cooperation under Anarchy.*

97. The term "zone of possible agreement" comes from Howard Raiffa's excellent *The Art and Science of Negotiation* (Cambridge, MA: Harvard University Press, 1982).

plants, far easier than verification of areal polluters, such as loggers or rice farmers. Previously regulated activities may already have data collection and dissemination systems on which a treaty can piggyback, much as the Long-Range Transboundary Air Pollution Convention has used economic data on fossil fuel usage to estimate emissions.[98]

Treaties usually provide for self-reporting by national governments, and the wide variance in levels of self-reporting suggests that some reporting systems work better than others.[99] Transparency can be increased by making rewards for compliance conditional on supplying such reports or allowing inspections. The human rights experience suggested to drafters of Agenda 21 that independent reporting by NGOs and industry had the ability and incentives to provide valuable compliance information.[100] On-site monitoring has been authorized in an international convention on wetlands as well as in nuclear weapons treaties.[101] Such measures skirt the self-incrimination problems involved in self-reporting and improve both the quantity and quality of data available. Treaty rules and procedures can also enhance information flow between parties, increase resources dedicated to monitoring, and finance the development of improved verification technologies.[102]

98. Marc Levy, "European Acid Rain: The Power of Tote-board Diplomacy," in Haas, Keohane, and Levy, eds., *Institutions for the Earth*.

99. United States General Accounting Office, *International Environment: International Agreements Are Not Well Monitored*, GAO/RCED-92-43 (Washington, DC: GPO, 1992).

100. Kathryn Sessions, *Institutionalizing the Earth Summit: The United Nations Commission on Sustainable Development* (New York: United Nations Association of the USA, 1992); and Jonathan Alexander Glass, "Selective Service: Non-Governmental Organizations' Influence in Enforcing International Environmental Agreements," unpublished undergraduate thesis, Harvard College, Cambridge, MA, March 1993.

101. *Convention on Wetlands of International Importance Especially as Waterfowl Habitat*, 2 February 1971, 996 U.N.T.S. 245, reprinted in 11 I.L.M. 969 (1972), 5 I.P.E. 2161, hereinafter cited as *Ramsar Convention*.

102. Finlayson and Zacher, "The GATT and the Regulation of Trade Barriers," 313. This has been an especially important contribution of the Standing Consultative Commission in the arms control arena; see Ralph Earle, "SALT II Compliance Controversies," in Michael Krepon and Mary Umberger, *Verification and Compliance* (Cambridge, MA: Ballinger Publishing Co., 1988).

For the data collected on behavior to lead to an appropriate response, treaty staff, governments, or NGOs must analyze and interpret the data to identify both the cases and sources of noncompliance. Often this may require a forum to inquire further into actions taken and the reasons behind them.[103] By increasing both the actors and the means authorized to collect, analyze, and disseminate information, a treaty can increase the likelihood that those tasks will be accomplished. By improving the ability to detect violations and the likelihood of detecting them, transparency fosters all parties' abilities to invoke reciprocity, sanctioning, and inducement strategies.[104]

Noncompliance Response System

If a treaty's compliance information system succeeds in developing information about noncompliance and noncompliers, successfully altering the noncompliers' behavior requires some response. Although responsive actions available to actors were discussed above, the following sections discuss how treaties can increase the likelihood that such responses occur and are effective.

Facilitating Compliance   Treaty organizations and secretariats can increase the likelihood and effectiveness of positive inducements that states can take unilaterally. They can provide mechanisms for sharing the burden of financing, making it feasible to fund projects that are larger than a single country would fund. These arrangements also establish a set of commitments and correlated expectations regarding contributions from various states to such a fund. While the International Monetary Fund and the World Bank have funding problems, nations in general are willing to make funds available for project funding. International organizations may also be able to fund projects that would be too controversial

---

103. See, for example, Robert Jervis, "From Balance to Concert," 73; Finlayson and Zacher, "The GATT and the Regulation of Trade Barriers," 298; and Stephen D. Krasner, "Regimes and the Limits of Realism: Regimes as Autonomous Variables," in Krasner, ed., *International Regimes*, 362.

104. See Jervis, "From Balance to Concert," 73; Oye, "Explaining Cooperation Under Anarchy," 15; and Fisher, *Improving Compliance with International Treaties*, 317.

for individual donor countries to fund. Whether such cooperative funding ventures will succeed in environmental affairs remains to be seen, as the Montreal Protocol and the Framework Convention on Climate Change are the first to include financial transfer mechanisms.

Facilitating compliance may also involve international education efforts targeted at clarifying rules and the means of complying. For example, Sweden and the International Maritime Organization created the World Maritime University to provide courses on marine pollution requirements and how to comply with them. Technology transfers also help. The Montreal Protocol requires signatories to provide one another with technological assistance to address noncompliance due to technical incapacity. The convention's highlighting of the fact that compliance by developing states depends upon these financial and technological transfers may help press developed states to fulfill these commitments.[105] Financial and technological transfers provide useful levers to induce compliance when sanctions may be politically difficult to impose, as among allies. Inducements themselves ease the noncompliance detection problem, since a government that might ignore a freestanding reporting requirement might well prove more forthcoming if providing the information opened up new funding sources.

These inducements can influence actors that want to comply but are unable, as well as providing incentives for countries to reexamine the costs of compliance and the priority given to it. Even states planning to violate may reconsider if the financial transfers are large enough. Inducements have their own problems, however. Joint funding itself has to overcome the inherent problem of free riders. Also, inducements cost more than sanctions because the former may require compliant actors to pay for compliance by actors that would have complied anyway, while successful threats and sanctions impose compliance costs on the target of the threat.[106] While international funding, technology, and education programs may prove difficult to enact and small in magnitude, they may nonetheless make compliance more likely.

---

105. *Montreal Protocol,* Article 5(5).

106. As Thomas Schelling has pointed out, promises are costly when they succeed; see *The Strategy of Conflict* (Cambridge, MA: Harvard University Press, 1960/1980).

Sanctioning Violation Proponents of sanctions contend that "compliance can be obtained efficiently by making violation unattractive rather than by altering the costs or benefits of compliance."[107] While sanctioning strategies face major constraints on their effectiveness in the international environment, their use remains a recurring theme in efforts to strengthen compliance systems.[108] As one example, Gro Harlem Brundtland, the Norwegian Prime Minister, has recommended ensuring compliance with carbon dioxide emission targets by establishing "an international authority with the power to verify actual emissions and to react with legal measures if there are violations of the rules."[109]

Sanctioning can be made more likely through several means. Regular international meetings of treaty parties increase the opportunities for actors to bring diplomatic and public pressures on noncompliant actors to change and explain their behavior.[110] Treaty rules that require publication and dissemination of information gathered from self-reported and independent sources provide a "basis for a wider critique and evaluation of a party's performance and policies" by countries, companies, and NGOs that have incentives to respond to noncompliance.[111] Media reports of noncompliance and pressures from domestic NGOs do not evoke questions of sovereignty that governmental sanctions often raise. NGOs may also be more likely than governments to impose political and economic costs on noncompliant governments and corporations. For example, Greenpeace and Sea Shepherd have frequently taken action to

107. Young, *Compliance and Public Authority*, 20.

108. See Fisher, *Improving Compliance with International Treaties*, for an excellent discussion of some of these limitations. The most extensive empirical study of international sanctions is Gary Clyde Hufbauer and Jeffrey J. Schott, *Economic Sanctions Reconsidered: History and Current Policy* (Washington, DC: Institute for International Economics, 1985). Keith Hawkins describes the limits of such strategies in the domestic realm in *Environment and Enforcement: Regulation and the Social Definition of Pollution* (Oxford: Clarendon Press, 1984).

109. Gro Harlem Brundtland, "The Road from Rio," *Technology Review* 96 (April 1993), 63.

110. Axelrod, *The Evolution of Cooperation*, 132; and Chayes and Chayes, "Adjustment and Compliance Processes in International Regulatory Regimes."

111. Chayes and Chayes, "Compliance without Enforcement," 323.

identify and prevent whaling.[112] By fostering such diffuse but influential pressures, improved transparency can increase compliance even absent formal sanctions.[113]

Treaties that formally authorize and "assign responsibility for applying sanctions" increase the expectation that "a given violation will be treated not as an isolated case but as one in a series of interrelated actions."[114] As Morgenthau notes, "nobody at all has the obligation to enforce" international law.[115] Few countries even have incentives to do so. To address the concerns over sovereignty that counteract a nation's desire to sanction another government for treaty violations, treaties can redefine what constitutes infringement of sovereignty by authorizing certain actions in the event of treaty violations.[116] Treaties can remove the international legal barriers, such as those stated in the General Agreement on Tariffs and Trade, that constrain those countries with incentives to enforce.

Finally, incentives to sanction can be increased by "privatizing" the benefits and costs of enforcement activity.[117] The 1911 Fur Seal Treaty between Japan, Canada, the United States, and the Soviet Union strictly curtailed killing seals at sea while leaving the United States and the Soviet Union in exclusive control of kills on the seals' breeding islands. The latter countries annually compensated Japan and Canada with a share of the kills in exchange for their halting their wasteful at-sea killing. The seal population recovered from three hundred thousand in 1911 to its preexploitation level of 2.5 million by the 1950s.[118]

112. See, for example, David Day, *The Whale War* (San Francisco: Sierra Club Books, 1987).

113. Young, "The Effectiveness of International Institutions," 176–178; and Chayes and Chayes, "Compliance without Enforcement," 8.

114. Axelrod and Keohane, "Achieving Cooperation under Anarchy," 234–237.

115. *Politics Among Nations*, 266.

116. Rules similar to those banning trade with nonparties to the *Montreal Protocol* could be applied to noncompliant parties (Art. 4).

117. Kenneth Oye, "The Sterling-dollar-franc Triangle: Monetary Diplomacy 1929–1937," in Oye, ed., *Cooperation under Anarchy*.

118. Simon Lyster, *International Wildlife Law: An Analysis of International Treaties Concerned with the Conservation of Wildlife* (Cambridge, England: Grotius Publications, 1985), 40–41.

Privatization can also involve authorizing governments to enforce treaty provisions directly against nationals of other countries caught violating treaty rules, as is seen in the provisions of the Convention on Trade in Endangered Species for penalizing noncompliers and confiscating specimens.[119] This "use of domestic enforcement procedures is likely to be possible in an increasing range of cases, like environmental treaties, where international regimes are aimed ultimately at influencing the private activities rather than state behavior."[120] By authorizing and establishing standards for second-level enforcement, treaties can increase its frequency. Governments balk less at sanctioning individual and corporate nationals of other governments than at sanctioning those governments directly. Whether such a strategy can work depends on whether states with enforcement incentives have effective jurisdiction over the regulated activity, but treaties with provisions that focus sanctions on individual actors provide one means of increasing compliance.

Preventing Violations   In addition to inducements and sanctions to elicit compliance, treaties, like domestic regulations, have a third strategy they can use. Some international standards rely on efforts to raise obstacles to and otherwise prevent noncompliance in the first instance. If primary rules can be established that succeed at coercing compliance in the first place, the demands placed on those responsible for monitoring and enforcing treaty regulations can be dramatically reduced. To implement such a strategy requires greater attention to "premonitory" control measures, that is, efforts to inspect and survey behavior before violations occur rather than to detect and investigate them afterwards.[121]

A coerced compliance system relies on finding regulatory chokepoints at which limits can be placed on the ability to violate a treaty's terms, thereby restraining actors that might otherwise be inclined to not comply. Careful selection of the point of regulatory intervention can allow treaties to target actors that have fewer incentives to violate rules or target

119. *CITES.*

120. Chayes and Chayes, "Compliance without Enforcement," 318.

121. Albert J. Reiss, Jr., "Consequences of Compliance and Deterrence Models of Law Enforcement for the Exercise of Police Discretion," *Law and Contemporary Problems* 47 (Fall 1984).

transactions between actors with differing incentives. Often it entails restricting the most transparent activity in the train of actions that precede an environmentally harmful action; preventing an activity that is not environmentally harmful itself may nevertheless provide the most effective means of preventing violations. Domestic examples include requiring manufacturers to install catalytic converters on automobiles or banning handgun sales.

A deterrence-oriented approach underlies the use of sanctions, which require the successful detection, prosecution, and sanctioning of violations after they have been committed to deter actors from committing them in the future. In contrast, a coerced compliance model greatly reduces the need for "postmonitory" control that can identify violations after they have been committed. Postmonitory control efforts face the common problems of detecting violations that violators desire to keep hidden; identifying the perpetrator of a detected violation, which may not always be obvious; collecting legally legitimate proof of the violator's having perpetrated the violation, and ensuring that the sanctions imposed are sufficient to deter others. A preventive orientation can be designed into treaty provisions by ensuring that the actors responsible for preventing violations have both the incentives and the authority to do so.

Time and the Treaty Process

Beyond the instrumental compliance policies involved in the compliance systems just discussed, treaties can reflect longer-term, less direct efforts to induce compliance. Treaty processes can seek to influence the ways governments and other actors perceive and define their interests so they will increasingly see compliance as furthering their interests. Treaty institutions can facilitate the creation of new information and knowledge and thereby change perceived interests.[122] Parties previously inclined not to comply may independently come to prefer to comply.

A treaty can authorize and fund scientific research and information sharing to investigate the benefits and costs of a state's own behavior to its independent self-interest. Compliance becomes more attractive if new

122. Alexander Wendt provides a detailed argument for the endogeneity of interests to international institutions in "Anarchy Is What States Make of It," *International Organization* 46 (Spring 1992).

knowledge shows that there are greater benefits associated with compliance, such as avoidance of the higher costs of environmental degradation. New information can increase domestic political pressures for compliance. The Long-Range Transboundary Air Pollution Convention provided scientific exchanges and a joint data base on transborder fluxes of pollutants that convinced some states to change their behavior.[123] Such international exchanges were essential in showing states the degree to which their environmental fate depended on the actions of their neighbors, increasing incentives for both compliance and enforcement. Research and development regarding cheaper means of complying provides a long-term solution to the problems of incapacity and high compliance costs as sources of violations.

Treaty processes can also encourage a process of social learning by which governments and other actors come to alter their values and objective functions. The negotiation and renegotiation process can change the value actors place on certain goals. Thus treaties can increase concern about the environment simply through promoting regular discussions that increase the understanding and perceived importance of the issue. This process involves more than reassessing costs or discovering new ways to achieve existing goals; it involves changing those goals. International organizations can consciously shape their perceptions of self-interest. For instance, developing and disseminating information regarding pollution and helping place scientists in domestic policy positions altered interests that increased compliance with treaty rules regulating Mediterranean pollution.[124]

A Framework for Analysis

The preceding discussion provides a foundation for identifying both whether and how treaties induce compliance. Most treaties have parties that combine the different types of motivations outlined. Even in what

123. Levy, "European Acid Rain."

124. When facing uncertainty, especially in environmental treaties, politicians may defer to scientific experts to define the nation's interests and implement the treaty domestically. These scientists have strong incentives to ensure their country complies with and enforces environmental treaties (Peter Haas, *Saving the Mediterranean*, 54–63).

appear to be collaboration problems, it seems likely that some states comply because they have an independent interest in complying, others are willing to comply if enough others do so, and others violate, complying only if they receive sufficient side payments for compliance and/or sanctions for violation. Some might act no differently if there were no treaty, while others may be complying only due to direct threats of sanctions. High compliance levels in general, and increases in compliance levels in particular, may be due to a treaty, but may be due to a range of other factors as well.

The wide array of sources of compliance that are exogenous to a treaty suggests the need for a healthy skepticism in suggesting that a treaty has caused compliance. More important, these sources of compliance provide rival hypotheses that may explain why compliance would increase at the same time as a change in rules. They provide a checklist of "likely suspects" that will help us avoid attributing causation to rules in cases in which other factors are responsible for changes in compliance.

Changes in the power and interests of dominant states may lead both to the adoption of more stringent rules and to better compliance. Treaty rules and correlated behaviors may often be merely different indicators of the same factors of power and interests that drive much of international politics. Not coincidence, but the existence of deeper causal links, will regularly cause changes in compliance levels to correlate with rule changes. Economic and technological changes might increase actors' incentives and capacities to comply, and domestic political changes might also lead actors to change their calculus regarding compliance. These factors may confound efforts to causally attribute a change in compliance to a particular rule change. Looking for specific independent evidence of such changes, however, allows us to confirm this analysis if appropriate evidence is found and to eliminate it if evidence shows that no such change has occurred. If changes in rules do appear to have caused increased compliance, however, the review presented here delineates the types of changes to the primary rule system, compliance information system, or noncompliance response system that we should expect to see.

This chapter has drawn on pragmatic concerns about the sources of treaty compliance and the current views of political scientists regarding the sources of international behavior to identify the major potential sources of compliance. The model of compliance proposed here highlights the role of both governments and private actors in all aspects of the compliance process. It also develops the notion that treaty compliance systems can be categorized as primary rule systems, compliance information systems, or noncompliance response systems. The fact that compliance can, at least theoretically, result from independent as well as treaty-related factors implies that analysis must not only show compliance correlating with differences in treaty rules, but also demonstrate that the treaty-related factors caused the compliance and other factors did not. This chapter has identified the means by which treaty rules may explain different compliance levels and provides a checklist of rival explanations that might explain a change in compliance more readily than the treaty rule. The following chapter provides a description of the political, economic, and diplomatic history of the improvements to the compliance system established for the conventions known as OILPOL and MARPOL. The theoretical model and empirical method of the current chapter provides the foundation for the analysis of these agreements that follows in chapters 4 through 8.

3

Intentional Oil Pollution: History and Context

For most people, the term *oil pollution*[1] conjures up images of massive oil spills from tanker accidents like those of the *Exxon Valdez*.[2] Yet intentional oil discharges from tanker operations have consistently over-shadowed accidents as the major source of ship-related oil pollution (see Table 3.1).[3] Oil is the pollutant with the longest history of international attention, having first become the subject of an international confer-ence in 1926. Though initial regulatory attempts failed, in 1954 nations signed the International Convention for the Prevention of Pollution of the Sea by Oil (known as OILPOL).[4] This first convention established

1. Portions of this chapter appear in Peter M. Haas, Robert O. Keohane, and Marc A. Levy, eds., *Institutions for the Earth: Sources of Effective International Environmental Protection* (Cambridge, MA: The MIT Press, 1993).

2. The *Exxon Valdez* spilled 35,000 tons of oil into Prince William Sound, Alaska, on 24 March 1989.

3. See National Academy of Sciences, *Petroleum in the Marine Environment* (Washington, DC: National Academy of Sciences, 1975); National Academy of Sciences and National Research Council, *Oil in the Sea: Inputs, Fates and Effects* (Washington, DC: National Academy Press, 1985); and MEPC 30/Inf. 13 (13 September 1990). All subsequent document citations refer to IMCO/IMO docu-ments, which are numbered according to issuing committee, meeting number, agenda item, and document number. Thus, MEPC 30/14/7 would indicate the seventh document issued relating to agenda item 14 of the thirtieth meeting of the Marine Environment Protection Committee. Information documents are des-ignated by "Inf." and the document number, e.g., MEPC 30/Inf. 13. Circulars are designated by "Circ." and the circular number, without a specific meeting number, e.g., MEPC/Circ. 138.

4. *International Convention for the Prevention of Pollution of the Sea by Oil,* 12 May 1954, 12 U.S.T. 2989, T.I.A.S. no. 4900, 327 U.N.T.S. 3, reprinted in 1 I.P.E. 332, hereinafter cited as *OILPOL 54; International Convention for the*

Table 3.1
Sources of oil input into the sea (million metric tons per year)

	Year of estimate		
	1971	1980	1989
Transportation			
Tanker operations	1.080	0.700	0.159
Dry-docking	0.250	0.030	0.004
Terminal operations	0.003	0.020	0.030
Bilge and fuel oils	0.500	0.300	0.253
Accidents	0.300	0.420	0.121
Scrappings	no est.	no est.	0.003
Subtotal	2.133	1.470	0.569
Offshore production	0.080	0.050	no est.
Municipal and industrial wastes and runoff	2.700	1.180	no est.
Natural sources	0.600	0.250	no est.
Atmosphere (emissions fallout)	0.600	0.300	no est.
Total	6.113	3.250	0.569
Discharge from tanker operations	1.080	0.700	0.159
Crude traded (mta)	1100	1319	1097
Discharge as % of crude trade	0.098	0.053	0.015

Sources: National Academy of Sciences, *Petroleum in the Marine Environment* (Washington, D.C.: National Academy of Sciences, 1975); National Academy of Sciences and National Research Council, *Oil in the Sea: Inputs, Fates and Effects* (Washington, D.C.: National Academy Press, 1985); and MEPC 30/INF.13 (19 September 1990).

a compliance system to regulate oil tanker discharges that went through extensive modification before being superseded by the International Convention for the Prevention of Pollution from Ships of 1973 and its 1978 protocol (known together as MARPOL 73/78).[5] Taken together, the numerous changes to the components of these agreements' compliance

Prevention of Pollution of the Sea by Oil, as Amended in 1962, 11 April 1962, 600 U.N.T.S. 332, reprinted in 1 I.P.E. 346, hereinafter cited as *OILPOL 54/62*; and *International Convention for the Prevention of Pollution of the Sea by Oil, as Amended in 1962 and 1969*, 21 October 1969, reprinted in 1 I.P.E. 366, hereinafter cited as *OILPOL 54/69*.

5. *International Convention for the Prevention of Pollution from Ships*, 2 November 1973, reprinted in 12 I.L.M. 1319 (1973), 2 I.P.E. 552; and *Protocol of 1978 relating to the International Convention for the Prevention of Pollution*

systems provide an excellent natural laboratory for testing hypotheses regarding the influence of rule changes on treaty compliance.

This chapter provides the contextual foundation for the later empirical analyses of government and industry responses to these compliance system changes. The material in this chapter bridges the theoretical framework of the previous chapter and the empirical chapters that follow. It identifies how the factors influencing compliance and compliance change (discussed in general terms in the previous chapter) have actually operated in the specific case of intentional oil pollution. The empirical chapters provide more in-depth analysis of the specific forms and motivations of each rule (the independent variable of interest), the exogenous factors that influenced compliance (the control variables), and the evidence of compliance change (the dependent variable). This chapter provides a broader overview of the problem, the relevant actors, and the history of the international regulation of intentional oil pollution.

The chapter describes the problem of intentional oil pollution[6], explaining the technical, economic, and political factors that produce it, including the structure of world oil trade. I detail the various operational and technological solutions that have been promulgated. I then describe the history of international oil pollution control since 1954, detailing the major changes that have been made since then in the treaty compliance system and in the larger technical, economic, and political contexts. This is followed by a delineation of the major actors involved in the bargaining over international oil pollution control and how changes in their interests and power have altered the nature of the international bargain over time. Taken as a whole, the chapter provides a general overview of both the major exogenous and endogenous factors contributing to compliance with international oil pollution regulations and how those have changed over time. This overview then provides the background against which the success of the specific efforts to improve compliance with oil pollution regulations can be examined.

---

from Ships, 17 February 1978, reprinted in 17 I.L.M. 1546 (1978), 19 I.P.E. 9451, hereinafter cited as *MARPOL 73/78.*

6. I will use the terms *intentional discharges* and *operational discharges* interchangeably, the point being to distinguish them from accidental oil spills.

Description of the Problem

In 1953 tankers transported some 250 million metric tons of oil by sea and intentionally discharged some 300,000 metric tons of it into the ocean.[7] Since then, the growing volume of crude oil transported by sea has resulted in estimates of total discharges that have ranged up to five million metric tons per year.[8] These discharges have traditionally represented two-thirds of all ship-generated oil pollution, with tanker accidents and discharges from nontankers making up the rest. Ships themselves generate only one-third of all oil entering the oceans, however, with land-based sources, natural seeps, and offshore production contributing the balance.

Why would a tanker captain intentionally discharge oil at sea? After a tanker delivers its cargo, a small fraction remains in the tanks as "clingage," adhering to the tank walls like the residue visible after a milk bottle has been emptied. Two standard industry practices led to the mixing of this clingage with seawater during return "ballast" voyages. The first involved tankers' filling empty cargo tanks with seawater as ballast to stabilize the tankers during their return voyages. The second involved tankers' use of seawater in high pressure cleaning machines to

7. The estimate adjusts James E. Moss's 1963 estimate for the smaller amount of oil transported in 1953. Moss estimated that, absent any control effort, 449 thousand metric tons of oil would be intentionally discharged given the 378 million metric tons transported in 1963; see *Character and Control of Sea Pollution by Oil* (Washington, DC: American Petroleum Institute, 1963), 51. Some 250 million tons of oil were transported by sea in 1953; see Sonia Zaide Pritchard, *Oil Pollution Control* (London: Croom Helm, 1987), 76.

8. The range of more recent estimates reflects both the increase in oil transported and the variance in assumptions underlying different analysts' estimates. For alternative estimates, see the National Academy of Sciences estimates noted in note 3; Arthur McKenzie, "Letter to the Honorable John L. Burton," in U.S. House of Representatives, Committee on Government Operations, *Oil Tanker Pollution—Hearings: July 18 and 19, 1978* (95th Congress, 2nd session) (Washington, DC: GPO, 1978); Sonia Zaide Pritchard, "Load on Top: From the Sublime to the Absurd," *Journal of Maritime Law and Commerce* 9 (1978); and J. Wardley-Smith, ed., *The Control of Oil Pollution* (London: Graham and Trotman Publishers, 1983).

wash down their tanks before receiving more oil.[9] Captains traditionally discharged the resultant oil-water mixtures (or "slops") at sea prior to arrival. While clingage represents only some 0.3 to 0.5 percent of total cargo, this translates into three hundred to five hundred tons of oil discharged for each voyage of a typical tanker. Given the vast volumes of oil transported by sea, such discharges can quickly accumulate into a major pollution problem.

Discharges of crude oil, while they do not persist indefinitely as believed by authorities in the United States and the United Kingdom in the 1920s, can remain afloat over long distances despite physical, chemical, and biological processes that tend to break them down over time.[10] Therefore, crude and other persistent oils discharged many miles offshore can subsequently appear on coastal beaches, posing both environmental and aesthetic threats. Their most visible environmental impact, and the most frequent source of public concern, has been their effect on seabirds. When birds come in contact with oil, it destroys the ability of their feathers to provide insulation and can cause internal damage when they ingest it in an attempt to clean themselves. Beyond this impact, however, scientists disagree over the extent of environmental harm caused by oil. Some scientists contend that the low-concentration but chronic and frequent oilings due to intentional discharges cause major long-term harm to fish, shellfish, and other forms of marine life.[11] Others find no evidence

9. Oil and lubricants can also leak into the bilges of tankers and other ships and mix with seawater. Although discharges of these mixtures also contribute to intentional oil pollution and have been regulated under OILPOL and MARPOL, tank cleaning and deballasting have constituted the major sources of oil from tankers and are therefore the major focus of the study undertaken here.

10. Refined oils evaporate quickly, causing relatively little marine pollution. Crude and fuel oils persist far longer in the marine environment and have constituted both the bulk of oil transported and the major focus of international regulation since the 1950s; see W. M. Kluss, "Prevention of Sea Pollution in Normal Tanker Operations," in Peter Hepple, ed., *Pollution Prevention* (London: Institute of Petroleum, 1968), 102–103.

11. Myriad studies describe the environmental impacts of oil pollution. For the most comprehensive collection of these estimates, see National Academy of Sciences and National Research Council, *Oil in the Sea*. For an excellent analysis of the impact of oil pollution on birds, see C. J. Camphuysen, "Beached Bird Surveys in the Netherlands 1915–1988: Seabird Mortality in the Southern North

that even major oil spills "have unalterably changed the world's oceans or marine resources."[12] Although less dramatic than major tanker spills, small oil patches and tar balls on resort beaches have also prompted regular public complaints, especially in developed countries. These environmental and aesthetic concerns have provided the impetus for essentially all the efforts to create international regulations governing oil pollution.[13]

The obstacles to such regulatory pressures include both technical and economic factors that create incentives to industry to discharge slops at sea as well as political, social, and legal factors that create obstacles to governments' incentives and ability to force industry to refrain from these discharges. Ocean oil pollution is both an externality and a commons problem. As with many industrial practices, tankers discharge oil at sea because it is the cheapest means of disposing of waste produced during oil transportation. Until the 1970s, the costs to oil transporters of devising new transportation processes to waste less oil exceeded the benefits. The value of the oil not wasted (dependent on oil market factors) did not

---

Sea since the Early Days of Oil Pollution," *Technisch Rapport Vogelbescherming* 1 (Amsterdam: Werkgroep Noordzee, 1989). For discussion of the early debate, see United Kingdom Ministry of Transport, *Report of the Committee on the Prevention of Pollution of the Sea by Oil* (London: Her Majesty's Stationery Office, 1953).

12. National Academy of Sciences and National Research Council, *Oil in the Sea,* 489. Even major accidental pollution from oil spills seems to have only limited, nonpermanent environmental effects. A 1990 study noted that oil "continues to be a matter of concern locally after accidents have released large amounts of oil that accumulate in sheltered areas, affecting amenity and living resources, especially bird life. While the damage is not irreversible, recovery can be slow"; see GESAMP (IMO/FAO/UNESCO/WMO/WHO/IAEA/UN/UNEP Joint Group of Experts on the Scientific Aspects of Marine Pollution), *The State of the Marine Environment,* GESAMP Reports and Studies no. 39 (New York: United Nations, 1990), 2. For other examples, see Second International Conference on the Protection of the North Sea, *Quality Status of the North Sea* (London: Her Majesty's Stationery Office, 1987), 72–73; United Kingdom Royal commission On Environmental Pollution, *Eighth Report: Oil Pollution of the Sea* (London: Her Majesty's Stationery Office, 1981), 38, 46–49, and 266; and J. M. Baker, *Impact of Oil Pollution on Living Resources* (Gland, Switzerland: International Union for Conservation of Nature and Natural Resources, 1983), 40.

13. See United Kingdom Ministry of Transport, *Report of the Committee,* 1–2.

outweigh the costs of the necessary recovery technologies and processes. The externality resulted from the fact that the transporter's calculus did not include the costs that oil pollution imposes on numerous but widely dispersed beachgoers, birdwatchers, and environmentalists. The collective harm is great, but each person harmed feels only a small share of it and has few incentives to organize to prevent it. Concentrated shipping and oil industries reap the benefits of discharging at sea in the form of reduced production costs, and consumers receive cheaper oil. In a domestic context, these factors would make control difficult to establish because of industry's greater incentives and ability to deploy resources against regulation.

The social and political context of oil pollution also make it an international commons problem. Resolution of such problems depends on agreeing to a level of governance that accounts for all social costs and benefits. Domestically, societies can opt for state control or privatize resources to resolve "how best to limit the use of natural resources so as to ensure their long-term economic viability."[14] Internationally, however, these options for restricting overuse and pollution of the ocean fall victim to the incentives that actors face in any commons problem and to governments' deference to the international norm of sovereignty.[15] Tankers discharge oil throughout the seas, causing most states to view the high costs of attempting to control discharges as greater than the benefits—in terms of pollution avoided—of doing so. States interested in creating effective regulations cannot do so without international action, because none has the legal authority or practical ability to control more than a small share of the transporters responsible for the problem.

International law categorizes nations as flag, coastal, or port states based on their relationship to a given vessel. A flag state is the state of registry, a coastal state is a state through whose territorial waters a vessel

---

14. Elinor Ostrom, *Governing the Commons: The Evolution of Institutions for Collective Action* (Cambridge, England: Cambridge University Press, 1990), 1. Ostrom conducts an excellent analysis of the resolution of common pool resource problems at the domestic level without government intervention.

15. See David Goetze, "Identifying Appropriate Institutions for Efficient Use of Common Pools," *Natural Resources Journal* 27 Winter (1987) on the differences between private goods, public goods, and common pool resources.

passes, and a port state is a state into whose ports a vessel enters. International law traditionally has strictly limited enforcement action by any state other than the flag state. Without a supranational government to enforce treaties, national governments enforce them by enforcing domestic laws that implement treaty provisions rather than enforcing the treaties directly. Under international law, flag states have essentially exclusive enforcement jurisdiction. They can monitor, investigate, prosecute, and penalize violations of their domestic laws by all tankers registered ("flagged") in them, no matter where such violations occur. Other states' enforcement rights are severely limited, however. If a tanker commits a violation in a state's territorial waters but does not subsequently enter any of its ports, that coastal state can only collect evidence of the violation for referral to the flag state, which retains enforcement jurisdiction. Only if a tanker violates international law in the territorial seas of a country and also enters one of that country's ports does that port state have internationally authorized jurisdiction to investigate, prosecute, and penalize the tanker.[16] Thus the coastal states that bear the brunt of the pollution that would create incentives for them to take action against tankers often face legal limits on their ability to do so.

While nations with incentives to control pollution lack the authority to do so, those with the authority may lack the incentives. For example, Liberia has historically been the registry of 15 to 30 percent of all tankers. But Liberia lies off major oil transportation routes and receives little coastal pollution from tankers. It therefore has few direct incentives to control discharges by companies providing it with major revenues since the citizens of other nations bear the costs of the pollution.

In short, oil pollution arises from economic and sociopolitical problems at the individual and national levels: oil transporters and oil con-

---

16. Since the U.N. Law of the Sea has been signed but not yet entered into force, the legal limits to the enforcement jurisdictions of coastal, port, and flag states are open to considerable interpretation. Many Law of the Sea provisions are becoming "customary international law" by incorporation into standard state practice or by declaration. However, those articles that directly address enforcement jurisdiction with respect to pollution from ships (Articles 217–220 and 228 of *United Nations Convention on the Law of the Sea*, 2 December 1982, UN Doc. A/Conf. 62/122, reprinted in 21 I.L.M. 1261 [1982]) have not been the focus of much national action or analysis.

sumers benefit from lower production costs, while beachgoers and bird-lovers suffer; governments and citizens of flag states benefit from higher revenues, while citizens of coastal states suffer. International norms and institutions do not provide the opportunities that can exist domestically to overcome the externalities and commons problems involved. Nations face a classic collaboration problem in which "each actor would rather share in such use of the resource that leads to depletion than to see its own restraint allow either the continued existence of the resource for others' use or the disappearance of the resource because the others show no restraint."[17]

The Basics of International Oil Pollution Regulation

Despite the numerous obstacles to regulation just delineated, nations have concluded numerous agreements requiring their governments to implement and enforce regulations that would reduce oil discharges. To facilitate the discussions of the history of oil pollution control and the actors and their incentives that appear later in this chapter, I next describe the basic legal control strategies that have been adopted and the various technologies and procedures by which tankers can control their intentional discharges.

The first approach adopted to address intentional discharges, whose use has been ongoing, involved delineating discharge "prohibition" zones and defining what discharges within and outside those zones are considered illegal. The philosophy underlying this approach has been that, if mixed with enough water and made far enough from those areas nations care about, discharges will dissipate sufficiently to mitigate their environmental harm. While the specifics have varied over time, OILPOL and MARPOL have coupled zones at fixed distances from the shorelines of all countries with areas deemed deserving of special protection. The rules could not prohibit all discharges since tankers needed, at a minimum, to discharge the oil-free water from their ballasting and tank-cleaning operations. The conventions therefore had to define what constituted

17. Arthur A. Stein, "Coordination and Collaboration: Regimes in an Anarchic World," in Stephen D. Krasner, ed., *International Regimes* (Ithaca, NY: Cornell University Press, 1983), 129.

"prohibited" discharges within and outside the specified zones. Inside these zones, limits were initially set on the maximum oil content for discharges measured in parts of oil per million parts of water. The rules were later changed to permit only "clean ballast," which was defined in terms of whether it would produce "visible traces of oil." Outside these zones, initially there were no limits, but subsequently limits were imposed governing the maximum rate of discharge measured in liters per mile.[18]

A second strategy adopted by both OILPOL and MARPOL has involved restricting total discharges per ballast voyage. This approach seeks to reduce rather than merely diluting and redistributing the oil entering the ocean. Since tank cleaning and deballasting both occurred on return trips from delivering oil, total discharge limits were defined as a maximum fraction of a tanker's total cargo. A third strategy has been to specify design and equipment standards that tankers must meet by specified dates depending on their size and date of construction. The technologies required reduce the amount of slops actually generated as well as the tankers' dependence on "the human element."

Tanker operators can control their discharges by using one or more of four different procedures and technologies. They can reduce the amount of oil-water mixtures generated during deballasting and tank cleaning. They can separate water from these slops that is pure enough to meet established standards if it is discharged. They can monitor discharges to ensure they meet the standard. And they can retain on board any oil-water mixtures that cannot be legally discharged and find a legal method of discharging them.

Industry has developed two different technologies to decrease the slops generated during oil transportation. Segregated ballast tanks (SBT) reduce slops created by the use of cargo tanks for ballast.[19] A tanker lacking SBT has a single piping system that feeds all the tanks on board, pumping oil in and out of all tanks at the beginning and end of each loaded voyage and pumping water in and out of the needed ballast tanks at the beginning and end of each ballast voyage. A tanker with SBT sets aside tanks and other ballast spaces equal to 30 to 40 percent of the

---

18. A discharge rate limit helped reduce a discharge's environmental impact by spreading the same amount of discharge over a greater distance.

19. William G. Waters, Trevor D. Heaver, and T. Verrier, *Oil Pollution from*

deadweight capacity of the tanker and outfits these tanks with a separate piping system. These tanks and connected pipes never carry oil, and the water in them can be discharged without monitoring. SBT tankers, however, must separate, monitor, and retain on board those slops resulting from loading additional ballast in cargo tanks during bad weather and from washing tanks for maintenance, residue control, and preparation for a cargo incompatible with the previous load.

Crude oil washing (COW) reduces the amount of oil wasted due to clingage. Originally proposed as an alternative to using SBT, COW addresses slops generated by tank cleaning, not ballasting.[20] COW substitutes crude oil for the water that traditional tank-washing systems used. Oil sprayed through the tanks during delivery acts as a solvent on and reduces clingage, increasing the total cargo delivered. Some clingage remains, however, so COW tankers that do not have SBT must still take ballast water into cargo tanks where it mixes with some oil. COW tankers must also undertake additional water cleaning in situations like those just described in the discussion of SBT tankers. Thus COW tankers must also separate, monitor, and retain slops.

MARPOL 73/78 required the use of both technologies. Each significantly reduced the total slops generated. A tanker of 100,000 deadweight tons that would generate 248 tons of oil without any pollution control would generate 168 tons using SBT and 100 tons using COW, but only 53 tons if the tanker used both technologies.[21]

A tanker can reduce the amount of oil that it discharges by effectively separating the oil from the water in the slops. Since the 1950s many tankers have done this by consolidating ballast and tank-cleaning slops in a single tank and letting gravity do the work.[22] Effective separation

---

Tanker Operations—Causes, Costs, Controls (Vancouver, BC: Center for Transportation Studies, 1980), 89–91.

20. Alan B. Sielen and Robert J. McManus, "IMCO and the Politics of Ship Pollution," in David A. Kay and Harold K. Jacobson, eds., *Environmental Protection: The International Dimension* (Totowa, NJ: Allanheld, Osmun & Co., 1983), 160.

21. Waters, Heaver, and Verrier, *Oil Pollution from Tanker Operations,* 120.

22. Oily water separating equipment was required for separating the oil from water in bilge residues of nontankers but has never proved a viable alternative for tankers because of the large volumes of throughput involved.

requires two to four days, a period usually available on ballast voyages.[23] After separation has been completed, a tanker can decant relatively pure water from underneath the oil. Any remaining slops can either be retained on board for discharge at port reception facilities or have new cargo loaded directly on top, systems known as retention on board (ROB) or load on top (LOT), respectively. These systems differ only in the means of slop disposal. Since using reception facilities often involves costly delays in port, LOT proves preferable since combining slops with the new cargo allows them to be discharged as part of the next delivery.[24]

Regardless of the quantity of slops generated or the means of separating oil out of slops, a tanker can reduce oil discharges by monitoring how much oil the slops contain. While implicitly required by the discharge standards of the OILPOL convention, oil discharge monitoring and control systems (ODMCS) were not required on tankers until MARPOL. An ODMCS measures and records the oil content, rate, and total quantity of the discharges taking place. It also automatically stops discharges that exceed the treaty limits.[25] However, reliable versions of these system have taken remarkably long to develop. Indeed, accurate oil content meters have only recently become available.[26] Crews therefore have needed to record ship speed and discharge volume to monitor the discharge rate and have had to visually monitor discharges to ensure that they did not violate the "clean ballast" rule.[27]

Tankers can reduce discharges by the methods they use for waste disposal. Tankers have essentially three legal methods of disposal available: discharge slops at sea within the prescribed limits on content and rate and constrained by the total discharge standards, discharge slops during delivery by using LOT, and discharge slops at reception facilities

23. Waters, Heaver, and Verrier, *Oil Pollution from Tanker Operations,* 76–77.

24. LOT cannot be used effectively, however, on short or bad-weather voyages or when the new cargo is incompatible with the previous one.

25. Although ODMCS devices provide an interesting approach to improving compliance, the absence of information on how many ships have installed them or the data that they record prevents their analysis at present.

26. See note 7 in chapter 7.

27. It was precisely technical problems with monitoring devices that provided the impetus for amendment so that crews could monitor their own discharges more readily.

after the ballast voyage. The latter approach places a corollary requirement on governments to ensure that reception facilities are available when the former two methods cannot be used. Reception facilities are most needed in oil-loading ports for tankers arriving after ballast voyages. At ports with reception facilities, discharging slops requires an additional three to nineteen hours in port, at considerable opportunity cost to the tanker.[28] When heading to ports without reception facilities, tankers can either use LOT, retain slops on board until they reach the next port with reception facilities, or discharge slops prior to arrival. Even though SBT, COW, and LOT have dramatically reduced the need for reception facilities, there are enough circumstances under which tankers have slops that cannot be legally discharged that reception facilities remain needed in many oil-loading ports.

The History of International Oil Pollution Regulation[29]

The first signed international agreement regulating intentional oil pollution was negotiated at a conference hosted by the British in 1954. this was not, however, the first time intentional oil discharges had been discussed internationally. Indeed, oil pollution had been on the international agenda since the 1920s. While oil pollution was not provoking widespread public concern at that time, nongovernmental organizations (NGOs) were vocal enough in the United Kingdom, the United States, and a few other countries to induce several nations to unilaterally restrict ship discharges near their ports.[30] Recognizing the international nature of the problem, the United States convened a governmental conference

28. Waters, Heaver and Verrier, *Oil Pollution from Tanker Operations*, 113. These authors estimate tanker time at an opportunity cost of $570 per hour in 1976 dollars.

29. This section relies heavily on Pritchard's exhaustive history of early (pre-1973) attempts at intentional oil pollution regulation in *Oil Pollution Control*. For the period since 1973, this section relies on the excellent analyses in R. Michael M'Gonigle and Mark W. Zacher, *Pollution, Politics, and International Law: Tankers at Sea* (Berkeley: University of California Press, 1979); and Sielen and McManus, "IMCO and the Politics of Ship Pollution."

30. Pritchard, *Oil Pollution Control*, 11 and 25–30, and United Kingdom Ministry of Transport, *Report of the Committee*, 3.

in 1926 that drafted a treaty to regulate intentional discharges. The United Kingdom provided the impetus for a similar effort in 1935 under the auspices of a League of Nations Committee of Experts. While both succeeded at drafting conventions, neither was ever signed or entered into force. However, the discussions undertaken during these efforts included issues such as the enforceability of zonal arrangements, how compliance with discharge provisions would be monitored and verified, the magnitude of penalties, reception facility requirements, and the distribution of enforcement jurisdiction between flag and coastal states, thereby foreshadowing debates that recurred throughout negotiation of the OILPOL agreement and its subsequent modification and later replacement by MARPOL.

The 1954 International Convention for the Prevention of Pollution of the Sea by Oil

Although efforts prior to World War II failed to achieve any signed agreements, the draft agreements shaped the negotiations that arose after the war. Growing demand for crude oil shipped from the Middle East but refined in western countries meant that more tankers were discharging the more persistent crude (versus refined) oils after tank-cleaning and ballast operations.[31] The number of complaints of spoiled beach resorts and large numbers of dead seabirds grew rapidly in the United Kingdom and elsewhere in Europe.[32] The United Kingdom continued to lead the call for international regulation. Particular interest groups rather than the public at large continued to be the major source of pressure for action by the United Kingdom. NGOs including bird protection societies, hotel and tourist organizations, and local governments banded together to form the U.K. Advisory Committee on the Prevention of Oil Pollution of the Sea (ACOPS).

In response to such pressures, the United Kingdom established the Committee on the Prevention of Pollution of the Sea by Oil, known as

31. J. H. Kirby, "The Clean Seas Code: A Practical Cure of Operational Pollution," in *International Conference on Oil Pollution of the Sea* (Rome, Italy: 1968), 203.

32. United Kingdom Ministry of Transport, *Report of the Committee*, 1.

the Faulkner Committee.[33] Its 1953 report concluded that the persistence of crude oils made the prewar zone system merely a "palliative." It recommended an international ban on discharges above 100 parts per million (ppm) by all ships (both tankers and nontankers) throughout the oceans. Until international agreement was reached, ships registered in the United Kingdom should be banned from discharges over 100 ppm "within a wide zone around the United Kingdom."[34] In contrast the United States, feeling it had solved its pollution problems through unilateral measures and voluntary compliance by industry with the zone system, had lost interest in international regulation.[35] As the United States delegate phrased it, "The pollution problem in the United States today is less critical than it was a quarter century ago, notwithstanding we are consuming three times as much petroleum and products today as we did in 1926."[36] Germany, the Netherlands, and other states—lacking domestic concern over coastal pollution, believing oil evaporated and biodegraded if discharged far from shore, or seeking to protect their maritime interests—saw regulation as unnecessary, but accepted zones as a means of placating the United Kingdom and averting unilateral action.

The British belief that zones were inadequate and the desire to avoid encumbering U.K. domestic shipping and oil interests led the United Kingdom to press for international action. The United Kingdom was also under pressure to do so from ACOPS, which held a con- ference in 1953, inviting environmental groups, national governments, and oil and shipping interests.[37] At the conference the British government announced that it would host the intergovernmental conference recommended in the Faulkner Committee's report.[38]

Thirty-two countries attended the governmental conference of 1954 in London. Although many European nations considered oil pollution a

33. United Kingdom Ministry of Transport, *Report of the Committee,* 1.

34. United Kingdom Ministry of Transport, *Report of the Committee,* 33.

35. Pritchard, *Oil Pollution Control,* 84.

36. International Conference on Pollution of the Sea by Oil, "General Committee: Minutes of 3rd Meeting Held on 30th April 1954," 4.

37. M'Gonigle and Zacher, *Pollution, Politics, and International Law,* 84.

38. Although IMCO—established under the United Nations in 1948—was the obvious forum for such negotiations, it did not start operation until 1958.

problem, most developing and Soviet bloc nations still did not.[39] Both the size and the biological effects of the oil pollution problem were debated.[40] The United Kingdom proposed to limit discharges throughout the ocean, essentially requiring all tankers to *stop* discharging waste oil at sea rather than merely discharging as far from shore as possible under a zonal approach.[41] Tankers, it was contended, could eliminate oil pollution if they "refrained from cleaning their cargo tanks and mixed oily ballast residues with new cargo oil" or practiced retention on board.[42] Oil and shipping companies objected because discharging slops at sea could be done while tankers were underway, while retention on board required lengthy port delays to discharge slops. Governments resisted the complementary requirement that their ports provide expensive reception facilities to receive these slops. Most countries felt such costs were unwarranted given that they themselves were not experiencing severe effects from oil pollution. The United Kingdom attempted to increase concern, even flying delegates to its beaches to demonstrate the problem, but "if domestic experience had little effect on states' policies, this predictably had even less."[43]

The failure to agree to limit discharges throughout the oceans left the final 1954 OILPOL agreement reflecting "the fact that most governments were still not willing to accept any important control costs themselves or

39. For particular states' views of the seriousness of oil pollution during the 1954 conference, see Pritchard, *Oil Pollution Control*, 98–99; and United Nations Secretariat, *Pollution of the Sea by Oil* (New York: United Nations, 1956).

40. For example, the Faulkner report itself concluded that there was no evidence that fish or shellfish beds were harmed by oil pollution (United Kingdom Ministry of Transport, *Report of the Committee*, 2–3). The French argued at the 1954 conference that their research had "produced no proof that its effects upon marine life were harmful" (International Conference on Pollution of the Sea by Oil, "General Committee: Minutes of 5th Meeting Held on 5 May 1954," 5). By this time, crude oil tankers clearly had become the major source of oil pollution.

41. M'Gonigle and Zacher, *Pollution, Politics, and International Law*, 90. While the final agreement addressed both tankers and non-tankers, the discussion here focuses exclusively on regulations relating to tankers, as they had become by far the major source of the problem.

42. Pritchard, *Oil Pollution Control*, 95.

43. M'Gonigle and Zacher, *Pollution, Politics, and International Law*, 87.

even to impose such costs on their industries."[44] The convention's primary rule system consisted of a prohibition against discharges above a specified limit within specified zones, defined as 100 ppm and fifty miles.[45] The convention imposed no restrictions on discharges outside the zones or on total discharges, thus relying on redistribution of discharges outside the zones to mitigate environmental damage. With respect to reception facilities, the final agreement required states to "ensure [their] provision" so as to meet the needs of nontankers, but it did not require equivalent measures for tankers, ensuring that tankers would have limited alternatives to discharging at sea.[46] Even these weak clauses regarding reception facilities led several countries to lodge objections.

Enforcement was based on language modified from the 1935 draft text, which required ship masters to record all ballasting, cleaning, and discharge operations in a newly developed oil record book.[47] Port states could inspect these books but not delay the ship and were limited to providing evidence to flag states for prosecution of violations.[48] All states were to ensure that penalties for violations on the high seas, that is, outside the territorial sea boundaries where they had exclusive prosecutorial jurisdiction, were equivalent to those within territorial seas. All states were required to report to the secretariat on reception facilities installed and application of the treaty, and flag states were required to report on actions taken on violations referred to them for prosecution. In 1958 the convention received the requisite ten ratifications, with five from major shipping states, and the first international rules regulating oil discharges entered into force.[49]

44. M'Gonigle and Zacher, *Pollution, Politics, and International Law,* 89.

45. Some wider zones were established near Australia and the North Sea states and in the Atlantic off the European and U.K. coasts.

46. Pritchard, *Oil Pollution Control,* 108.

47. This had been recommended in the Faulkner report as an amendment to the Oil Pollution Act of 1922 as well; see United Kingdom Ministry of Transport, *Report of the Committee,* 32.

48. Pritchard, *Oil Pollution Control,* 112.

49. "Major shipping states" were defined as those having an aggregate tonnage of over 500,000 tons.

This initial compliance system did not look promising. Since no existing monitoring devices could reliably measure the 100 ppm that was the standard set for discharges, even conscientious captains could not assure compliance except by making all discharges outside the specified zones. Many captains could have done this at little additional cost in time or fuel, since those going from Europe to the Middle East could deballast and clean their tanks in the still legal discharge area in the central Mediterranean.[50] Yet whatever the cost, the benefits of compliance were all but nil, since the definition of compliance and the inherent difficulty of monitoring tanker behavior in fifty-mile coastal zones made successful detection of violations extremely implausible. Likewise, the evidentiary and incentive-related obstacles posed by giving flag states exclusive jurisdiction made prosecution, let alone penalization, appear unlikely.[51] In 1961, after three years of operation, the Intergovernmental Maritime Consultative Organization (IMCO) summarized responses from the contracting parties, noting that they "indicate the Convention is obviously not perfect, since it is the first international instrument attempting to control oil pollution, but that it has created a framework for further progress."[52]

The 1962 Amendments

In the years after signature of the 1954 agreement, the amount of oil transported by sea (and discharged at sea) increased rapidly. Concern over pollution increased, especially among states surrounding the Mediterranean. Dissatisfied with the results of OILPOL, ACOPS sponsored a conference of eleven countries in 1959 that recommended extending the 1954 zones and globally banning operational discharges.[53] This conference helped provide the impetus for IMCO to prepare the amending conference that parties to the 1954 conference had urged after a few years of experience with the rules. IMCO sponsored the conference in 1962, which thirty-eight states attended. Essentially a new scientific consensus

---

50. Kirby, "The Clean Seas Code," 203.

51. For a description of how these problems had been noted as early as the 1926 conference, see Pritchard, *Oil Pollution Control,* 23.

52. OP/CONF/2 (1 September 1961).

53. Pritchard, *Oil Pollution Control,* 119.

had emerged, prompted by French and German studies showing that crude and fuel oils persisted long enough that zones would not prove effective over the long term.[54] While they rejected the ACOPS recommendation that called for the establishment of wider coastal zones globally, several countries gained extended zones including the whole of the North and Baltic Seas.[55] Discharges below 100 ppm remained legal within these zones. Industry raised few objections to these extensions; whether the zones were fifty or one hundred miles, experience had shown that enforcement never extended beyond a country's three-mile limit.

By 1962 the British had modified their 1954 proposal to make it more acceptable: they proposed that only *new* tankers over 20,000 tons be banned from discharging anywhere in the oceans. The rule implicitly required new tankers to practice retention on board and monitor all discharges wherever they occurred, although oil content monitors and oily-water separators were not explicitly required. The United States strenuously objected to this proposal because, as tanker companies admitted, devices and procedures that would have allowed a tanker to monitor its own compliance with the oil content standard did not exist.[56] The United States complained that the primary rules had defined compliance in terms inconsistent with existing monitoring capabilities. To make compliance with the global discharge prohibition possible would also have required that ports provide reception facilities to receive slops generated. The parties did expand the scope of the 1954 reception facility provisions to include oil-loading ports, but they weakened the substantive requirement so that governments need only "promote" their provision. Despite opposition from the United States along with Japan, Norway, and the Netherlands, the proposed restriction on new tankers was adopted.[57]

Changes in enforcement and reporting were also considered. The parties adopted a clause from the 1935 agreement stating that penalties were

54. Pritchard, *Oil Pollution Control,* 130–131.

55. See Annex A of *OILPOL 54/62.* An intriguing French proposal to increase the speed of obtaining signatures by *decreasing* the size of prohibition zones off nonparty states failed (Pritchard, *Oil Pollution Control,* 133). This effort at reciprocity received little serious consideration.

56. Pritchard, *Oil Pollution Control,* 138.

57. Pritchard, *Oil Pollution Control,* 139.

to be severe enough to discourage violations. They considered but rejected proposals to increase the enforcement jurisdiction of coastal and port states and to require coastal states referring alleged violations to flag states to forward the information to IMCO. The conference recommended, but did not require, that IMCO "produce reports for which the Contracting Governments should contribute information" on oil pollution, convention effectiveness, reception facilities, enforcement, and violations.[58] Provision was also made to allow future amendments to be made within the IMCO structure rather than requiring a conference, laying the groundwork for the 1969 amendments.[59] The 1962 Amendments entered into force in 1967.

As late as 1975, a British oil pollution expert did not think "there was a tanker over 20,000 [tons] in the world complying with the 1962 Amendments."[60] There was no significant increase in reception facilities. However, the rules did result in increased research into alternative means of reducing discharges: the United States developed segregated ballast tanks, the Soviet Union developed chemical washing techniques, and British oil companies developed LOT.[61]

Oil companies recognized that expensive equipment would be required for new ships to comply with the general prohibition and that explicit requirement of the needed oily water separators and oil content meters was in the offing. Shell developed LOT after direct prompting by the British government.[62] While this prompting only required the company

58. Resolution 15 in IMCO, *Resolutions Adopted by the International Conference on Prevention of Pollution of the Sea by Oil, 1962* (London: IMCO, 1962).

59. Article XVI, *OILPOL 54/62.*

60. M'Gonigle and Zacher, *Pollution, Politics, and International Law,* 99.

61. Pritchard, *Oil Pollution Control,* 145. As James Kirby noted, "It was only our close study of the solution recommended by the [1962] conference of discharging oil ashore into slop facilities that really drove us towards the load-on-top method"; see J. H. Kirby, "Background to Progress," *The Shell Magazine* 45 (January 1965), 26.

62. Largely based on two articles by James Kirby of Shell, it has been suggested that the load-on-top technique was not developed until 1962; see Kirby, "Background to Progress"; and J. H. Kirby, "The Clean Seas Code." Shell does appear to have reevaluated the feasibility of combining slops with fresh cargo as a direct response to the requirement of the 1962 amendments that slops be dis-

to "discover" that "most crude oil cargos are compatible," thus removing Shell's major objection to mixing residues from one delivery with subsequent cargoes, it took the threat of international rules requiring expensive equipment to induce Shell to make the discovery.[63] By reducing the frequency of tank cleaning, consolidating slops via retention on board techniques, and discharging slops with the next load of cargo, oil companies could reduce the amount of equipment required and the amount of discharge time spent at reception facilities while decreasing the amount of oil lost due to clingage.

Many governments liked LOT because it reduced the total oil discharges and the need to build reception facilities. However, since oil companies designed LOT to avoid equipment costs, it required tanker operators to determine visually when to stop discharging water from beneath oil slops. Oil companies admitted that, in practice, this would frequently result in discharges exceeding 100 ppm by large amounts, violating the 1954 and 1962 discharge limits.[64] Nonetheless, by 1964 Shell had allegedly encouraged LOT adoption by some 60 percent of tankers, including most American and European ships.[65]

The 1969 Amendments

Since the 1920s, pressure to reduce oil pollution had rested on the belief that "once the stuff is in the sea, it is there for ever."[66] By the late 1960s,

---

charged ashore (Kirby, "Background to Progress," 26.) However, the process of consolidating tank-cleaning and deballasting slops was described in the 1953 Faulkner report; see United Kingdom Ministry of Transport, *Report of the Committee*. Moss states that by 1963 most U.S. tankers were already mixing crude oil slops "with the next cargo"—the major innovation of load on top—as part of normal procedure (Moss, *Character and Control*, 42).

63. Kirby, "The Clean Seas Code," 206.

64. Accurate oil content meters were still not yet available.

65. Kirby, "The Clean Seas Code"; and M'Gonigle and Zacher, *Pollution, Politics, and International Law*, 97.

66. Sir Gilmour Jenkins in Kirby, "Background to Progress," 26. See also the U.S. Bureau of Mines study provided to the 1926 conference and the Report on the Second Session of the League of Nations Committee of Experts of October 26, 1935, both cited in United Kingdom Ministry of Transport, *Report of the Committee*, 6–9, as well as the Ministry of Transport conclusions themselves.

however, the evidence was "overwhelming" that natural processes made oil "unobjectionable" over time.[67] Nonetheless, by then it had become "axiomatic that the less oil discharge into the sea, the better."[68] In this context, the grounding of the *Torrey Canyon* in 1967 provided a major new impetus to oil pollution control. The accident raised public concern in many European countries, and major international agreements to address tanker accidents were quickly signed. Growing environmentalism was also raising broader concerns over all ocean pollution.

ACOPS again hosted a conference that helped push operational discharges onto the international agenda. Its Rome conference of 1968 occasioned a major proposal by Shell to scrap the existing zonal system and the implicit equipment requirements of the 1962 amendments to OILPOL in favor of promoting voluntary adoption of LOT, or what Shell called the Clean Seas Code.[69] The issue of modifying international regulations to legitimize LOT and eliminating equipment requirements had already been raised in IMCO's newly established Subcommittee on Oil Pollution (SCOP) in 1965. The United Kingdom, which had been the major force for oil pollution reduction up until the 1962 conference, now began working much more closely with its oil companies. In 1968, it proposed to the IMCO subcommittee that all governments promote LOT. At the same time, growing domestic environmentalism was leading the United States to seek stronger international controls.[70]

These conflicting efforts to modify the convention came to a head in the subcommittee during 1968. On one side oil and shipping companies, supported by the United Kingdom, Norway, the Netherlands, and France, sought to have LOT legitimized to avoid the costs of the equipment required by the 1962 amendments. Indeed, as late as 1967 "an effective oil-monitoring system still had not been developed and oily-water separators were not always effective."[71] To make the LOT procedure legal

67. C. T. Sutton, "The Problem of Preventing Pollution of the Sea by Oil," *BP Magazine* 14 (Winter 1964), 9. This fact did not address, however, the environmental damage that oil could cause before it became "unobjectionable."

68. Kirby, "The Clean Seas Code," 210.

69. Kirby, "The Clean Seas Code."

70. M'Gonigle and Zacher, *Pollution, Politics, and International Law,* 100.

71. M'Gonigle and Zacher, *Pollution, Politics, and International Law,* 99.

required abandoning definitions of discharges in terms of oil content (ppm). To prove LOT's value, oil companies conducted joint research with the British government which showed that, by controlling the rate rather than the oil content of a discharge, essentially equivalent environmental protection could be achieved.[72] A rate metric, that is, volume discharged over a given distance, had the advantage that tankers already had the machinery necessary to assess compliance on board. The oil companies also sought reduction in the prohibition zones from fifty to fifteen miles.

On the other side, however, environmental states led by the United States were seeking to strengthen the discharge regulations. After much oil industry lobbying, the opponents of LOT were willing to accept the legitimation of LOT. They eliminated the 1962 requirement that had imposed equipment-dependent standards on new tankers. However, in exchange they required the industry to accept more stringent standards overall. The final amendments constituted a compromise. The fifty-mile zones were retained. Outside the zones, all tankers would need to keep discharges below a new limit defined in terms of rate (60 liters per mile, or 60 l/m) rather than oil content.[73] Within the zones, only discharges of "clean ballast"—those leaving no visible trace—would be allowed. Therefore, "any sighting of a discharge from a tanker . . . would be much more likely to be evidence of a contravention."[74] Finally, the United States seized on the oil industry's claim that LOT could ensure that the convention was "automatically enforced world wide": the United States successfully pressed for a provision that total discharges be limited to 1/15,000th of a tanker's cargo capacity. Even absent precise measurements, these rules allowed port authorities to assume that any tanker with clean tanks had blatantly violated the agreement.[75]

72. M'Gonigle and Zacher, *Pollution, Politics, and International Law,* 99.

73. The 60 l/m rate posed few problems for tankers since it was "a figure within which any responsibly run ship, no matter how big, could operate" (Kirby, "The Clean Seas Code," 208).

74. Resolution A.391 (X) (1 December 1977), Annex, par. 5.

75. Kirby, "The Clean Seas Code," 200 and 209; and William T. Burke, Richard Legatski, and William W. Woodhead, *National and International Law Enforcement in the Ocean* (Seattle: University of Washington Press, 1975), 129. Imagine a new tanker that loads 150,000 metric tons of oil in Kuwait. It delivers 149,400

The clean ballast, the rate metric, and the total discharge limits would all improve the primary rule system by increasing the ability, if not the incentives, of tanker operators to monitor their own compliance. The clean ballast and total discharge limits would, in addition, improve the compliance information system by making independent detection of violations possible.[76] The final bargain involved environmental states' agreeing to redefine the primary rule on discharges so that compliance would not require the installation of expensive equipment, in exchange for oil companies' agreeing to standards that would be more enforceable and would reduce oil pollution if complied with. This bargain represented significant movement toward definition of discharge standards in terms that corresponded to the existing capacities of oil companies to monitor their own behavior and of governments to verify that behavior.

The IMCO Assembly adopted these amendments in October 1969, dramatically changing the underlying principle of oil pollution regulation. The 1926, 1935, 1954, and 1962 rules had all permitted discharges except in prohibited zones. In contrast, the new rules prohibited discharges except under certain conditions.[77] And for the first time international rules required that the amount of oil entering the oceans be reduced rather than merely redistributed. However, ratifications were so slow that the 1969 amendments entered into force only in 1978.

---

metric tons in Rotterdam, 600 metric tons (0.4 percent) remaining as "clingage" of oil to the tanks' sides. On its return voyage to Kuwait, it ballasts several tanks with sea water and cleans others with sea water. It allows the oil to separate from the resulting oil-water mixtures and discharges the water overboard. If it arrives in Kuwait with less than 590 metric tons of oil residues ("slops"), it would clearly have discharged more than 1/15,000th of its 150,000 tons. The more likely scenario would involve arrival in Kuwait with completely clean tanks or negligible slops.

76. The concomitant changes to the compliance information system that would have made these changes effective were not made, however. International law barred port states from conducting the intrusive inspections necessary to verify compliance with the total discharge standards, and no changes were made to the flag state enforcement prerogative.

77. Samir Mankabady, *The International Maritime Organization: Volume I: International Shipping Rules* (London: Croom Helm, 1986), 318.

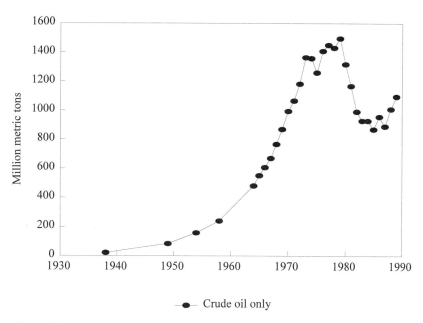

Figure 3.1
World seaborne oil trade, 1938–1989
Source: British Petroleum, *BP Statistical Review of World Energy* (London: British Petroleum Co., 1990).

The 1973 International Convention for the Prevention of Pollution from Ships

The growing environmental interest of the late 1960s became manifest in the early 1970s with the United Nations Conference on the Human Environment and the London Dumping Convention. Concern over oil pollution increased as seaborne oil trade increased from 158 million metric tons in 1954 to 1366 million metric tons in 1973 (see figure 3.1). Even if each tanker had discharged significantly less cargo than previously, the problem would have increased. Countries such as Greece and Italy that had previously opposed strict regulations adopted more environmentalist stances as they experienced more operational pollution and more calls for environmentalism at home.[78]

These forces had their strongest impact in the United States, which

78. M'Gonigle and Zacher, *Pollution, Politics, and International Law,* 118.

pushed for regulations stricter than the 1969 amendments. Although the LOT system had been legally legitimized only in 1969, oil companies had allegedly been using it since 1964. The United States believed this experience proved that it was far less effective than the oil companies alleged. The United States was especially concerned regarding the ease with which tanker crews could violate and the massive resources and diligence needed to detect violations.[79] Therefore, in the early 1970s the United States proposed supplementing the existing performance standards with equipment standards. These included requirements for existing tankers and ports to install the equipment necessary to comply with the 1969 performance standards, such as oil discharge monitors, oily water separators, dedicated slop tanks, and reception facilities. The new standards also required new tankers of over 70,000 tons to install SBT. In addition, to reduce spills during accidents, the United States proposed that new ships have double bottoms. The last two proposals were especially expensive.

The domestic pressures behind the international efforts of the United States also found expression in Congress' 1972 passage of the Ports and Waterways Safety Act. It required the Coast Guard to adopt strict unilateral equipment standards by 1976 unless other countries agreed to rules similar to those the United States was proposing. The Coast Guard was also to deny entry to any ships violating such rules.[80]

At the same time, Canada began an "aggressive diplomatic campaign" for protection of coastal states' environmental rights.[81] Motivated by both environmental and territorial concerns, Canada worked with other developed coastal states such as Australia and New Zealand to persuade developing states to attend the 1973 International Conference on Marine Pollution. This conference and the International Convention for the Prevention of Pollution from Ships (MARPOL) that it produced were far broader in scope than previous agreements. They applied to oil platforms

79. M'Gonigle and Zacher, *Pollution, Politics, and International Law,* 108.

80. U.S. Public Law 92–340, *Ports and Waterways Safety Act of 1972,* 10 July 1972, Sec. 201(13).

81. Jesper Grolin, "Environmental Hegemony, Maritime Community, and the Problem of Oil Tanker Pollution," in Michael A. Morris, ed., *North-South Perspectives on Marine Policy* (Boulder: Westview Press, 1988), 32.

as well as ships, included refined as well as crude oil, and used five annexes (including Annex 1 for oil) to address liquid chemicals, harmful packaged substances, sewage, and garbage discharged by ships. This broader perspective on pollution had been foreshadowed by the Intergovernmental Maritime Consultative Organization's renaming of the Subcommittee on Oil Pollution as the Subcommittee on Marine Pollution in 1969 and by the creation of the Marine Environment Protection Committee in 1973 as a full committee answering directly to the IMCO Assembly.

The U.S. proposals provided the basis for most of the conference's discussion on oil tankers. However, despite U.S. pressure, the final agreement maintained essentially the same performance standards that had existed before the conference.[82] The zones remained 50 miles, though special areas were designated in the Mediterranean, Baltic, Black, and Red Seas and in the Persian Gulf, but not the North Sea.[83] Outside the zones, discharges below 60 l/m remained legal (see table 3.2). Inside the zones, the negotiators provided a dual definition of "clean ballast": no "visible trace," as under the 1969 Amendments, or less than 15 ppm, which represented a reversion to an oil content metric.[84] The total discharge limit remained 1/15,000th of the cargo capacity for existing tankers, although new tankers were limited to 1/30,000th.[85] Within the special areas, reception facilities were to be required on a fixed schedule independent of the date of the treaty's entry into force, while outside these areas the requirement would become operative a year after its

82. The United States had sought measures beyond those needed to implement the 1969 amendments. It sought to widen the prohibition zones to one hundred miles while restricting allowable discharges within them to 10ppm and to reduce rate limits outside the zones from 60 to 30 l/m (M'Gonigle and Zacher, *Pollution, Politics, and International Law,* 113).

83. The special areas in the Red Sea and Persian Gulf were designated as part of the 1973 Convention. However, only the 50 mile zones applied until sufficient states had provided reception facilities at all oil loading terminals (Annex I, Regulation 10, *MARPOL 73/78*).

84. The U.S. was seeking a 10 ppm definition (M'Gonigle and Zacher, *Pollution, Politics, and International Law,* 113).

85. Annex I, Regulation 9, *MARPOL 73/78*.

Table 3.2
Intentional oil pollution discharge standards (1954 through present)

Convention date (dates in force)	Age of ship	Discharge limit within zones*	Discharge limit outside zones*	Maximum total discharge
1954 (1958–1967)	All	< 100 ppm	None	None
1962 (1967–1978)	Existing	< 100 ppm	None	None
	New	< 100 ppm	< 100 ppm	None
1969 (1978–1983)	All	Clean ballast	< 60 l/m	< 1/15,000 tcc
1973/1978 (1983–present)	Existing	< 15 ppm	< 60 l/m	< 1/15,000 tcc
	New	< 15 ppm	< 60 l/m	< 1/30,000 tcc

* = zones of 50 miles plus special areas.
ppm = parts oil per million parts water.
tcc = total cargo capacity.
l/m = liters per mile.

entry into force. The conference also agreed for the first time to incorporate a requirement for annual reporting on penalties and enforcement "in a form standardized by the Organization."[86]

The far more controversial aspects of the regulations adopted by the conference involved the equipment standards. The final standards required equipment to ensure compliance with the discharge standards—oily water separators and monitoring devices—to be placed on all new tankers delivered after 1979[87] and on existing tankers starting three years after the treaty entered into force. The requirement for segregated ballast tanks on new tankers initially evoked strong opposition from states with large shipping interests and from oil companies. Two factors reduced resistance to the SBT requirement: it promised to defuse unilateral U.S. adoption of the even more expensive double bottoms, and the recent construction boom meant the cost of building new tankers with SBT was far off in the future. Although the oil industry had initially opposed the

86. Article 11(f), *MARPOL 73/78*.

87. "New" tankers were defined as tankers for which building contracts were drawn up after 31 December 1975, whose keels were laid after 30 June 1976, or whose delivery occurred after 31 December 1979 (Annex I, Regulation 1, Par. 6, *MARPOL 73/78*).

SBT requirement, "with the American submission, the handwriting was on the wall, and the oil companies began intensive negotiations to consider its adoption."[88]

Large increases in the amount of oil transported by sea and a corresponding increase in discharges prompted new concern in many coastal states. Developed states with long coastlines and small shipping industries—like Australia, Canada, and New Zealand—supported the U.S. SBT proposal. Italy, traditionally opposed to stringent requirements, joined the environmental ranks as it experienced increased levels of coastal pollution. Developing states such as Egypt, Argentina, and India lent their support, as they faced growing levels of pollution from developed countries' ships and saw few direct costs to their own small-sized tanker fleets. They also supported the requirement of SBT as a means of reducing the amount of oil waste generated, thereby deflecting growing pressures to require them to build expensive reception facilities. In an era of détente, Soviet bloc countries saw support as having low economic costs and both political and environmental benefits.[89] This diverse coalition was large enough to pass the requirement.

In contrast, countries that opposed the SBT requirement included countries like Denmark, Germany, Greece, Norway, and Sweden which had large independent shipowning interests that were less capable than oil company-owned fleets of passing on the costs involved, and those like France and Japan which had shipbuilding interests that were concerned that new requirements would cause deferrals of ship orders.[90] The seven major oil companies initially resisted the SBT requirement, but with five based in the United States, they realized that they would face either strict unilateral rules or less stringent international rules. Through the oil company representative, the Oil Companies' International Marine Forum (OCIMF), they eventually supported the 1973 SBT rule. Independent

---

88. M'Gonigle and Zacher, *Pollution, Politics, and International Law,* 109.

89. For an insightful argument for the impact of détente on Soviet bloc positions in negotiations on acid rain, see Marc Levy, "European Acid Rain: The Power of Tote-board Diplomacy," in Haas, Keohane, and Levy, *Institutions for the Earth.*

90. M'Gonigle and Zacher, *Pollution, Politics, and International Law,* 114.

Table 3.3
MARPOL 73/78 SBT and COW requirements for large crude oil carriers
(tankers over 70,000 deadweight tons)

	Tankers delivered before 31 December 1979	Tankers delivered between 1 January 1980 and 30 May 1982	Tankers delivered after 1 June 1982
MARPOL requirement	SBT or COW	SBT only	SBT and COW

Source: Adapted from Y. Sasamura, "Oil in the Marine Environment," in *IMAS 90: Marine Technology and the Environment* (London: Institute of Marine Engineers, 1990), 3–4.

shippers through their representative, the International Chamber of Shipping, opposed the SBT requirement to the end, contending that entry into force and enforcement of the 1969 amendments would "effectively eliminate oil pollution arising from operational discharge."[91] Table 3.3 details the final equipment requirements. Finally, while states were required to "ensure provision" of reception facilities in all tanker ports, this wording left ambiguous whether states or industry was responsible for constructing them.[92]

The 1973 conference also sought to improve implementation, enforcement, and compliance. Continuing ratification delays were addressed through a tacit acceptance procedure that permitted entry into force of certain amendments unless more than one-third of the signatories explicitly objected. The conferees also applied construction standards to ships built after set dates, regardless of the number of ratifications.[93] Compliance with the equipment standards was to be established by initial surveys conducted by national governments and ship classification societies and documented in an International Oil Pollution Prevention (IOPP) certificate. States were given expanded rights to inspect the IOPP certificates of ships entering their ports and to determine whether they met the equipment requirements. If a ship was found in violation, govern-

91. MP XIII/2(c)/7 (2 June 1972), 2. See also IMP/CONF/8/4 (29 June 1973).

92. M'Gonigle and Zacher, *Pollution, Politics, and International Law,* 1 14–120.

93. This built on the approach of amendments adopted in 1971 that limited tank size to reduce the impact of accidental spills (Pritchard, *Oil Pollution Control,* 159).

ments were obligated to "take such steps as will ensure that the ship shall not sail until it can proceed to sea without presenting an unreasonable threat of harm to the marine environment."[94] Negotiators hoped that providing more environmentalist port states with such enforcement powers would improve compliance. While much more drastic increases in port state and coastal state enforcement powers were discussed during the conference, they were defeated due to the political power of the major flag states coupled with the desire of many states to make these jurisdictional decisions in the Law of the Sea context.[95]

The 1973 MARPOL convention included significant changes in the compliance system. The United States interpreted the low compliance rates and the difficulties of enforcing the discharge standards as evidence of the inherent flaws in such primary rules that would prevent the development of an effective compliance system. The conference's acceptance of an SBT requirement involved a new primary rule system that created a fundamentally different regulatory structure. These standards shifted responsibility for compliance from tanker captains to tanker owners. The rule also shifted the site of potential violation from the open ocean to the shipyard. This primary rule shift alone created a significant change in the compliance information and noncompliance response systems. However, the negotiators went further and also directly improved these systems. Given the new equipment standard, provisions explicitly provided new rights that allowed the compliance information system to "piggyback" on the existing information infrastructures of classification societies and government inspections of tanker certificates and tankers' actual conditions. The new convention also provided specific rights of detention that posed a significant deterrent threat should any country choose to use them. It also made more modest improvements in the reception facility requirements and reporting requirements that were essentially based on the notion that greater specificity about what and when obligations were to be met would increase compliance.

94. Article 5(2), *MARPOL 73/78.*

95. M'Gonigle and Zacher, *Pollution, Politics, and International Law,* 231–234.

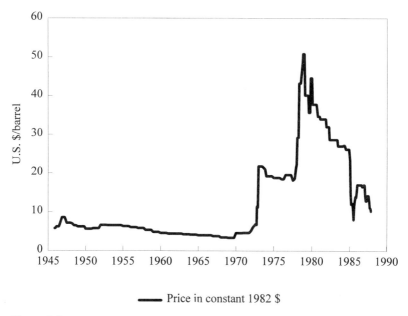

Figure 3.2
Crude oil prices
Source: Gilbert Jenkins, *Oil Economist's Handbook* (New York: Applied Science Publishers, 1990).

A final factor influencing oil pollution control during this period was the dramatic increase in oil prices during the 1970s. Before the first oil crisis in 1973, oil prices had remained essentially constant since 1950 (see figure 3.2). The shock of that crisis caused the value of the oil that tankers were discharging at sea to triple overnight, causing the benefit of using oil waste reduction and retention technologies like LOT, SBT, and COW to jump dramatically. Whether a company would adopt one of these technologies would depend on whether the oil retained at the new price of oil now outweighed the technology's cost.

The 1978 Protocol to the 1973 Convention
The 1973 MARPOL Convention failed to gain quick ratification both because of the strong resistance generated by the new equipment and reception facility requirements and because the adoption of Annex I, which addressed oil pollution, was legally linked to the adoption of the

more contentious and expensive Annex II, which addressed chemical pollution. Then, just as the *Torrey Canyon* incident had motivated earlier agreements, a series of accidents in December 1976 and January 1977 combined with activist pressures, including lawsuits by the Center for Law and Social Policy,[96] to produce unilateral U.S. action and put oil pollution back on the international agenda.

In response to the accidents and congressional pressure, the Carter Initiatives were proposed. These included the requirement of double bottoms and other systems to prevent accidental spills, but also addressed operational pollution by requiring that the Coast Guard unilaterally require SBT on all tankers over 20,000 tons and annual tanker inspections unless international negotiations produced stronger standards. Under direct threats that "if IMCO tailors its moves to suit and protect the U.S., we will accept; if not, we reserve the right to impose our own rules," IMCO called the Tanker Safety and Pollution Prevention Conference in 1978.[97] This conference produced a protocol that became integral to the 1973 MARPOL agreement; together they are known as MARPOL 73/78.

At the 1978 conference the United States proposed to expand the application of the 1973 SBT rule from new tankers over 70,000 tons to new and existing tankers over 20,000 tons. Various alternatives were proposed. The most important of these, pressed by the oil industry and the United Kingdom, involved requiring existing tankers over 70,000 tons to install COW. As had been the case with LOT in the 1960s, the oil industry reevaluated its technological options and COW, available since the late 1960s, became far more attractive in light of both rising oil prices and U.S. pressures to require SBT. Especially for cargo owners, such as oil companies, COW was far more attractive than SBT because it had lower capital and operational costs and reduced the waste of increasingly valuable cargo. To address the concerns over accidental pollution that had led to the expensive U.S. proposal of requiring double

96. Attorneys from the Center for Law and Social Policy, representing 15 environmental groups, were among the U.S. delegates to the 1973 and 1978 conferences; see Clifton E. Curtis, "Statement," in U.S. Senate, Committee on Foreign Relations, *Hearings on Protocol of 1978 Relating to the International Convention for the Prevention of Pollution from Ships, with Annexes and Protocols: June 12, 1980* (96th Congress, 2nd session) (Washington, DC: GPO, 1980), 9.

97. M'Gonigle and Zacher, *Pollution, Politics, and International Law,* 130.

bottoms, opponents proposed placing segregated ballast tanks in protective locations.

The support that in 1973 had greeted the U.S. proposal of requiring SBT on large, new ships evaporated with the 1978 proposal of expanding the requirement to all ships. The response to this proposal, and the attendant costs of retrofitting, revealed how little countries were willing to pay for environmental protection. A few states with very heavy pollution supported the SBT retrofit proposal. Support also came from states with large independent fleets, which had many tankers laid up during the tanker oversupply caused by the decreased demand for oil at post-1973 prices. Norway, Sweden, and Greece saw requiring SBT retrofits as a way of reducing each existing tanker's (and the fleet's) cargo capacity, which would allow their laid-up tankers to reenter the world market.[98] Most states, however, including Soviet bloc and developing ones, saw the SBT requirement as too costly, preferring the cheaper COW. The oil and shipping industries, and countries representing them like the United Kingdom, also preferred COW.[99] Given the "power and determination" of the United States, supporters of the COW alternative recognized the need for compromise.[100] In contrast to the debates of the late 1960s, even the United Kingdom no longer trusted exclusive reliance on performance standards. The final compromise required new crude tankers over 20,000 tons to install SBT and COW, while requiring existing tankers over 40,000 tons to install either SBT or COW.[101] Existing tankers were sure to choose the cheaper COW option.

While the protocol left performance standards unchanged, it added language requiring regular unscheduled inspections to verify compliance. The IOPP certificate and oil record book were modified. In addition, to speed entry into force the negotiators delinked ratification of Annex I on oil pollution from Annex II on chemicals. As with the 1973 rules,

98. M'Gonigle and Zacher, *Pollution, Politics, and International Law,* 123 and 135.

99. As discussed below, COW actually resulted in net savings for oil companies.

100. M'Gonigle and Zacher, *Pollution, Politics, and International Law,* 138.

101. "New" tankers under the protocol were defined as tankers for which building contracts were drawn up 1 June 1979, whose keels were laid after 1 January 1980, or whose delivery occurred after 1 June 1982 (Annex I, Regulation 1, par. 26, *MARPOL 73/78*). Existing tankers, instead of retrofitting SBT

applying the equipment requirements to ships delivered after June 1982 (regardless of the date of the new rule's entry into force) removed incentives for countries to delay ratification to slow the rule's impact. Ratifications from the requisite fifteen states, which represented not less than half the world's merchant tonnage, led to entry into force of the combined MARPOL 73/78 in 1983.

Essentially, the 1978 protocol merely added a new technology (COW) to the compliance system already established under the original 1973 MARPOL convention. However, the 1978 conference provides good insight into the strength of states' commitments to reducing oil pollution. The United States sought to extend the SBT requirements established in 1973 to essentially all tankers rather than applying them merely to large new ones. Most states, regardless of level of development or other factors, opposed the U.S. position as too costly, with some forty states preferring COW, four undecided, and only eleven favoring a requirement to retrofit tankers with SBT.[102] Most states clearly had limits as to how far they were willing to go to prevent oil pollution and to impose the costs of pollution prevention onto their industries. Maritime states and the industries they represented continued to have strong incentives to avoid standards that imposed costs on those industries, suggesting that they also had strong incentives not to implement and enforce existing agreements with vigor.

The 1980s

Since stringent regulations now existed on paper, some nations and the former Intergovernmental Maritime Consultative Organization (IMCO), now renamed the International Maritime Organization (IMO), sought to redirect their focus toward compliance. By 1981 the frequent changes in the proliferation of oil pollution regulations and the problem of adopting new regulations before old ones entered into force was inhibiting

---

or COW, could dedicate certain tanks to ballast for an interim period until 1985 for tankers over 70,000 tons and until 1987 for tankers over 40,000 tons. For a chart of the application of the various regulations, see Y. Sasamura, "Oil in the Marine Environment," in *IMAS 90: Marine Technology and the Environment* (London: Institute of Marine Engineers, 1990), 3–4.

102. M'Gonigle and Zacher, *Pollution, Politics, and International Law*, 136–137.

compliance. These factors led the IMO Assembly to resolve that the Marine Environment Protection Committee should consider amendments only "on the basis of clear and well-documented compelling need."[103] While this resolution has not prevented amendments, they have been largely technical in nature. All have been adopted through the MEPC and have entered into force via the tacit acceptance procedure of the 1973 convention.

The long delay in the ratification of MARPOL 73/78 meant that several amendments to the agreement were proposed before it even entered into force. These amendments were agreed to during regular meetings of the Marine Environment Protection Committee in 1981, 1982, and 1983 and were adopted in September 1984.[104] The changes were made to improve the existing equipment requirements or, at the request of shipping interests, to remedy implementation problems that had become evident through initial experience with MARPOL 73/78. They improved the specifications for the MARPOL oil monitoring, separating, and filtering equipment whose details had plagued IMO for years, waived equipment requirements under strict conditions, and again modified the oil record book. Since 1984 IMO has adopted several other amendments to MARPOL. In 1987 and 1990, respectively, the Gulf of Aden and Antarctica were designated as special areas deserving greater environmental protection. In 1990 the guidelines for MARPOL surveys were harmonized with those for other IMO conventions.[105] In 1991 and 1992 U.S. pressures in the wake of the *Exxon Valdez* spill resulted in amendments requiring new tankers to be built with double hulls or equivalent spill protection construction and all tankers to carry plans for dealing with any oil pollution emergency.[106]

103. Resolution A.500(XII) (20 November 1981).

104. Anonymous, "First Amendments to MARPOL 73/78 Adopted," *IMO News* 4 (1984), 8.

105. IMO, *Status of Multilateral Conventions and Instruments in Respect of Which the International Maritime Organization or Its Secretary-General Performs Depositary or Other Functions, as at 31 December 1990* (London: IMO, 1991).

106. Anonymous, "6 July Deadline for New Tankers," *IMO News* 2 (1993), 1; and Anonymous, "1991 MARPOL Amendments Enter into Force," *IMO News* 2 (1993), 2.

After 1978 enforcement concerns also grew in Europe. The *Amoco Cadiz* spilled 223,000 tons of oil off France on 16 March 1978. The spill prompted the Commission of the European Communities to start working on directives on the enforcement of oil pollution regulations and France to call a 1980 conference that led the maritime authorities of fourteen European states to adopt a Memorandum of Understanding (MOU) on Port State Control.[107] The MOU promulgated no new primary rules, and for ten years exclusively addressed inspections to monitor equipment and certificate violations. Only reluctantly in 1992 did member states amend the MOU to include inspections to monitor compliance with operational discharge requirements.[108] The purpose of the agreement was to improve the enforcement of IMO agreements on many maritime issues, oil pollution being just one. It is essentially a "nested" regime that forms one component of MARPOL's compliance information system and noncompliance response system. It requires each member state to inspect 25 percent of the ships entering its ports and to report certificate and equipment deficiencies as well as detentions to a central computer processing facility. Inspectors in each country can thereby

107. *Memorandum of Understanding on Port State Control in Implementing Agreements on Maritime Safety and Protection of the Marine Environment,* 26 January 1982, reprinted in 21 I.L.M. 1 (1982), I.P.E. II/A/26-01-82, hereinafter cited as *MOU.* Member states include Belgium, Denmark, Finland, France, Federal Republic of Germany, Greece, Ireland, Italy, Netherlands, Norway, Portugal, Spain, Sweden, and the United Kingdom. The 1982 MOU replaced a similar MOU signed in the Hague by eight North Sea states in 1978; see Secretariat of the Memorandum of Understanding on Port State Control, *The Memorandum of Understanding on Port State Control,* information pamphlet (The Hague: The Netherlands Government Printing Office, 1989). For a full discussion of the MOU, its history, and its operation, see George Kasoulides, "Paris Memorandum of Understanding: A Regional Regime of Enforcement," *International Journal of Estuarine and Coastal Law* 5 (February 1990).

108. In 1989 member states felt that "control on compliance with operational [discharge] requirements did not strictly fit in the present framework of the MOU"; see Secretariat of the Memorandum of Understanding on Port State Control's *Annual Report, 1989* (The Hague: The Netherlands Government Printing Office, 1989), 16. Member states used "extreme caution" in moving toward inclusion of discharge standards in MOU inspections; see Secretariat, *Annual Report, 1992* (The Hague: The Netherlands Government Printing Office, 1992), 21.

access recent data on violations by any ship arriving in a port.[109] The MOU represented a major change in the enforcement structure for the countries involved. Those drafting it sought to resolve the classic collaboration problem posed by enforcement among a set of actors with relatively equal power: the costs of enforcement accrue to the enforcing state, while the benefits accrue to the whole of Europe. The MOU strategy has involved developing a compliance information system to process and disseminate tanker data with sufficient speed and accuracy that all enforcement authorities have incentives to keep the system accurate and up to date. This process, in turn, has the potential to improve enforcement and thereby increase compliance with MARPOL's rules.

The European MOU has begun to influence port state enforcement in many other countries. In November 1992 the maritime authorities of ten Latin American countries signed an MOU that was in Spanish but was otherwise almost identical to the European MOU.[110] In 1992 Poland became the first new member since the MOU had been adopted. Russia and Croatia have applied for membership. Canada, Japan, and the United States have "cooperating maritime authority" status with the MOU and maintain their own rigorous enforcement programs and data collection systems.[111] IMO is fostering the development of similar agreements elsewhere, especially in the Asia-Pacific region.[112]

109. See George Kasoulides, "Paris Memorandum of Understanding: Six Years of Regional Enforcement," *Marine Pollution Bulletin* 20 (June 1989); and Kasoulides, "Paris Memorandum of Understanding: A Regional Regime of Enforcement," for extensive discussions of the Memorandum of Understanding on Port State Control.

110. *Acuerdo de Viña del Mar: Acuerdo Latinoamericano sobre Control de Buques por el Estado Rector del Puerto,* 5 November 1992. Original signatories include Argentina, Brazil, Chile, Colombia, Ecuador, Mexico, Panama, Peru, Uruguay, and Venezuela.

111. Secretariat of the Memorandum of Understanding on Port State Control, *Annual Report, 1992* 10–14.

112. Secretariat of the Memorandum of Understanding on Port State Control, *Annual Report, 1992,* 13–14; Anonymous, "1994 Target for Port State Control Pact," *IMO News* 2 (1993), 24; and Resolution A.682(17) (November 1991).

Actors, Interests, and International Bargains

The foregoing history of international efforts to regulate oil pollution allows us to examine the factors influencing parties to the international bargain and the obstacles that have impeded international regulation. Although governments have negotiated the international agreements and subsequent compliance system modifications that represent the independent variables of this study, the positions governments have taken and the paths by which these positions have become international law have been the product of three other sets of actors: domestic and international environmental constituencies, the oil transportation industries, and the international organizations that have engaged in the negotiations. The negotiations have been between people and NGOs in a few activist states pushing for increased control to reduce oil pollution's effects on birds and resorts, the affected industries lobbying nationally and internationally to avoid the costs of regulation, the governments that translate these pressures into national positions, and the international bodies that establish the context in which and the rules by which actors' positions become international law.

Publics and Nongovernmental Organizations

Two factors have prompted the calls for governments to take international action: activism by nongovernmental organizations (NGOs) and dramatic tanker accidents. Domestic NGOs, most notably British NGOs like the Royal Society for the Protection of Birds and the Advisory Committee on Oil Pollution of the Sea (ACOPS) in the 1950s and 1960s, and a U.S. NGO, the Center for Law and Social Policy in the 1970s, have provided the major impetus for the two states that have taken major leadership positions on oil pollution control. In contrast to their impact in other environmental issue areas, in the area of oil pollution environmental NGOs have had little direct input at the international level.[113] There are two exceptions. ACOPS is an international NGO that during

113. For example, the International Union for the Conservation of Nature (IUCN) has played an essential role in the development of the Convention on International Trade in Endangered Species and the Convention on Wetlands of International Importance. Greenpeace and other NGOs have played important roles in addressing whaling regulations under the International Convention for

its early years worked domestically in the U.K. while also sponsoring several conferences prior to major intergovernmental conferences that focused international attention on the issue of oil pollution and prompted proposals for treaty amendment. More recently ACOPS has become significantly less active on the issue. Although Friends of the Earth International acquired consultative status with the Intergovernmental Maritime Consultative Organization (IMCO) in 1973, it has only recently gained the expertise and legitimacy among other IMCO members that are necessary for it to begin to influence the organization.[114] Greenpeace acquired consultative status only in 1992.

Accidents have also played a major role in mobilizing public calls for action. The 1967 *Torrey Canyon* disaster, along with a general growth in environmentalism, strengthened and have sustained public concern and calls for international action in Europe and the United States, leading to both the 1969 OILPOL Amendments and the 1973 MARPOL Conference. Accidents in 1976 and 1977 prompted the Carter Initiatives and the 1978 conference. The *Amoco Cadiz* accident in 1978 prompted the European Community directives and the MOU agreement. Notably, major accidents off the coasts of developing states have not led those states to push for greater levels of pollution control.[115] Accidents have prompted public concern only among developed, democratic nations. Only these countries have both a process for expressing that concern and have addressed more immediate human welfare concerns sufficiently that environmental concerns could become a priority. Even among these states, concern has been by no means quick in arising, nor has it been universal. Until the 1970s most governments viewed oil pollution as a problem that did not really warrant an international solution.

Whether prompted by NGOs or by accidents, public concern rarely has been manifested as specific concern over noncompliance. It has been

---

the Regulation of Whaling. For an interesting account of the latter, read David Day, *The Whale War* (San Francisco: Sierra Club Books, 1987).

114. M'Gonigle and Zacher, *Pollution, Politics, and International Law,* 67; and interviews with Gerard Peet, the FOEI representative, in 1991 and 1992.

115. For example, despite numerous major tanker spills in the Persian Gulf region, only one of the eight states bordering the Gulf (Oman) is a party to MARPOL.

a few governments that have translated concerns motivated by major accidental spills, continuing pollution, and growing environmentalism into the various proposals since 1954 to improve the compliance system that regulates intentional discharges. Although NGOs and publics have focused much attention on major tanker spills and called for international regulations, neither group has been active in identifying or responding to violations of discharge or equipment standards.

The Oil Transportation Industries

Oil companies, shipping companies, and classification societies have all played major roles in developing and implementing international regulations through lobbying of national governments and through direct representation to IMCO. Oil companies have been directly influenced by the regulation of oil pollution in their capacities as both tanker owners and cargo owners. Oil companies, mainly the "seven majors" based in the United States and the United Kingdom, own almost one-third of all tankers and control even more through subsidiary corporations and long-term chartering arrangements. Independents, based mainly in Norway, Sweden, Denmark, and Greece, own the other two-thirds of the tanker fleet. They carry oil as cargo under charter to the major oil companies and others and work the "spot" market. Most owners register ships in their native countries and, with few exceptions, owners of both oil companies and independent tankers are concentrated in developed states.[116] Liberia and Panama have become the two major developing states to have significant tanker registries, but most ship owners are not Liberian or Panamanian nationals. Many owners register tankers in other nations, called flag-of-convenience states, to avoid stringent labor and equipment standards and high registration fees. For example, in 1975 U.S. companies registered over 80 percent of their tankers outside the United States, allowing the United States to exhibit the peculiar combination of significant ship ownership with strong environmentalism.[117]

As tanker owners and operators, oil companies—like independents—have interests in opposing regulations that increase construction and

116. M'Gonigle and Zacher, *Pollution, Politics, and International Law*, 56–58.
117. M'Gonigle and Zacher, *Pollution, Politics, and International Law*, 265 and 288.

operation costs or reduce their range of choices among oil transportation processes and technologies. Major oil companies have consistently sought to have regulations provide maximum flexibility, preferring voluntary guidelines and various equipment options to mandatory requirements. They have tended to support regulation more than independents, however, because they are more vulnerable to the pressures of unilateral regulation by the United States and the United Kingdom and of attacks on their high-visibility public images.[118] International regulation provides two benefits to companies threatened with unilateral regulation. First, intergovernmental negotiation generally produces less stringent rules than the activist state is seeking.[119] Second, internationalism eliminates the competitive disadvantage entailed by unilateral rules. Therefore, companies in activist states have tended to resist rules until unilateral regulation seemed inevitable, and then supported international action while urging the adoption of the least restrictive and least costly rules.

As cargo owners, oil companies have also sought to avoid international regulation that increased the costs of transporting oil by tanker or decreased the flexibility they had in such transportation. Therefore they have consistently resisted requirements of SBT retrofits (which decrease existing tanker capacity) and reception facilities (which require costly delays in port) as expensive and unnecessary. However, they have supported international regulations requiring LOT and COW, which allowed them indirectly to force the independents from which they chartered tankers to waste less oil during cargo voyages.

While oil companies have economic incentives to reduce discharges independent of environmental concerns, the same is not true for independents. Discharging oil imposes no costs on independents because they rarely own the cargoes they transport and they are paid for cargo loaded, not cargo delivered. In addition, all methods of avoiding such discharges involve equipment costs, operational costs, or both, and these costs are

118. The 1991 and 1992 television advertising campaign by Conoco proclaiming its progressive environmental stance based on its installation of double hulls prior to the dates required by the U.S. government provides one example of this. An early reference to the role of public reputation as a force for reducing oil pollution can be found in Moss, *Character and Control*, 46.

119. The rules are not always lowest common denominators, however, and can prove more stringent than laggard states would like.

borne by the tanker owners. Not surprisingly, therefore, independents, companies in laggard states, and their respective governments have tended to resist regulation.

Oil and shipping interests have been most manifest in coordinating domestic-level lobbying to influence the positions that governments bring to oil pollution negotiations. Tanker owners have done this through the International Chamber of Shipping (ICS) and, more recently, through the Independent Tanker Owners Association, known as INTERTANKO. Oil companies, originally represented through ICS, formed the Oil Companies' International Marine Forum (OCIMF) in 1970. Through this organization an oil company could "secure representation of its interests as *both* a shipowner and a cargo-owner."[120] Besides participating in domestic-level activism, these organizations have consultative status with IMCO and a wealth of technical expertise that allow them to exert considerable influence on the international regulatory process. They have also taken an interest in monitoring the adequacy of reception facilities in port and the use of LOT, and their monitoring activities serve as nongovernmental sources of compliance information.

While classification societies are represented with IMO through the International Association of Classification Societies, their influence tends to be greatest in the area of implementation. Classification societies, such as Lloyds' of London and Det Norske Veritas, regularly inspect commercial vessels and classify them based on factors such as their design, equipment, and condition. A vessel's level of classification plays an important role in the price and availability of insurance to the owner. As described at length in the chapter on industry responses, the presence of classification society representatives in shipyards provides a major means by which tanker owners are made aware of international equipment regulations and forced to comply with them.

Governments

The preceding discussion has delineated the factors influencing public and industry positions on oil pollution control. Governments mediate between these interests and international negotiation on oil pollution as well as bringing their own concerns over sovereignty and broader

---

120. M'Gonigle and Zacher, *Pollution, Politics, and International Law,* 65.

jurisdictional issues to bear. Several factors determine a government's position on oil pollution control in general and on the compliance system supporting it in specific.

Domestic pressure for stringent regulation depends on both the susceptibility of a nation's shoreline to oil pollution and the level of domestic concern. The former depends on both the size of the country's coastline and the location of that coastline relative to the major oil transportation routes. The latter, as already discussed is a function of NGO activism, recent tanker accidents, and general concern about the environment.

The size and political strength of domestic interests in the tanker-owning and tanker-using industries influence whether public and NGO pressures translate into government support for strong environmental rules. The success of oil and shipowning companies in getting governments to protect their interests depends on their power relative to public concern. These factors not only control the positions governments take in international negotiations, but also influence whether the government has an interest in effectively implementing treaties already in force and exercising control over its industries. Oil and shipping interests may wield sufficient economic and political power in some states, especially developing ones, to counterbalance other pressures on governments to take strong environmental stances both in negotiations and in enforcement.

A nation's position vis-à-vis the oil transportation trade—as an importer versus an exporter or as a coastal state versus a port state—can place it in a materially different position than it might otherwise have with respect to both current diplomatic proposals and implementation strategies. The burdens of enforcing rules on limiting total discharges during ballast voyages and supplying reception facilities fall almost exclusively upon oil-exporting states, since they have the ports where tankers arrive after their ballast voyages. In contrast, procedures related to equipment, certification, and discharge record keeping can be controlled in either loading or delivery ports. Coastal states along major trade routes may find themselves facing significant oil pollution, but lacking any ability to control the ships responsible.

A nation's level of development dramatically influences its government's preference for stringent regulation, or lack of it, in two ways. Developing states lack strong environmental constituencies pushing for environmental control because other welfare issues assume greater national priority. Developing states also lack the resources to dedicate to compliance and enforcement activities such as the provision of reception facilities or the monitoring and prosecution of discharge violations.

Finally, legal concerns over sovereignty have also influenced states' positions on compliance issues. These concerns were especially marked when governments viewed the resolution of enforcement jurisdiction issues in the context of MARPOL as influencing larger jurisdictional issues, especially those involved in the Law of the Sea negotiations. Many developing coastal states wanted to increase their power by expanding the enforcement jurisdiction of coastal states since "many more ships would pass through their two hundred-mile economic zone than would ever enter their ports."[121] Some developed states proposed allowing port states to prosecute violations that occurred outside their territorial seas. Maritime states, seeking to protect their shipping interests (and navies) from foreign interference, opposed proposals to limit the enforcement jurisdiction of flag states and to increase the boarding and seizure rights of coastal states on grounds that such regulations would hinder their freedom of navigation. The final MARPOL bargain avoided these contentious issues: it reiterated all states' rights to enforce the convention rules within their jurisdiction, but left the definition of jurisdictional limits to the Law of the Sea.

With respect to monitoring and enforcement, domestic public concern provides a major condition necessary for any state to monitor and enforce rules to limit intentional discharges. States with low levels of domestic concern over oil pollution, even if pressures during negotiation can garner their support for stringent pollution controls, do not establish and consistently fund the coast guards and marine police needed to effectively monitor and enforce such controls. Even states with high levels of public concern must overcome the pressures that any interested oil or shipping concerns bring to bear in opposing such activities. Flag-of-convenience states deriving large revenues from registries may have few incentives to monitor and enforce regulations. States may also lack the financial resources and administrative capacity to effectively implement

---

121. M'Gonigle and Zacher, *Pollution, Politics, and International Law,* 232.

the terms of an agreement.[122] Although political realities in developing states often place environmental concerns below other elements of social welfare, even if such concerns do develop, governments often lack the resources and personnel needed to carry out aerial surveillance programs or to follow through on prosecutions of violations referred by other states.

International Organizations

OILPOL was the product of a diplomatic conference in 1954. Initially undertaken by the British, secretariat duties were handed off to IMCO upon its establishment in 1958. IMCO is a specialized U.N. agency that has a large mandate to address all international shipping issues including safety, working conditions, load lines, and other standards for all types of vessels involved in international movement. IMCO has used diplomatic conferences (1962, 1973, and 1978) and has also established ongoing fora for evaluating existing rules and negotiating revisions to them. The Subcommittee on Oil Pollution and its successor, the Subcommittee on Marine Pollution, recommended treaty amendments through the Maritime Safety Committee and the IMCO Council to the IMCO Assembly. The assembly had the power to adopt amendments and "recommend" them to governments for ratification.

In 1973 IMCO increased the status of the marine pollution problem by creating the Marine Environment Protection Committee (MEPC), a full committee of IMCO that meets every eight months.[123] More important, MARPOL incorporated a "tacit acceptance" provision that gave the MEPC authority to adopt technical treaty amendments that would automatically take effect unless more than one-third of the parties objected.[124] Since MARPOL took effect, tacit acceptance has become the primary means used to revise oil pollution regulations.

Several international organizations have been established at the regional level to deal with responses to large oil pollution incidents and

122. See Marc A. Levy, Robert O. Keohane, and Peter M. Haas, "Improving the Effectiveness of International Environmental Institutions," in *Institutions for the Earth*.

123. Resolution A.297(VIII), (1973); and MEPC XII/13/3 (5 November 1979).

124. Article 16(2)(f) and 16(2)(g), *MARPOL 73/78*.

other matters affecting the marine environment, most notably the United Nations Environment Programme's Regional Seas Programmes. Some of these organizations have conducted studies on the provision of reception facilities in their region.[125] However, no international organization other than IMO, has sought to promulgate any regulations on intentional discharges. The group that drafted the MOU on Port State Control, described above, is the only other international body that has taken any significant interest in intentional discharges.

Negotiating a Bargain

Prior to the 1970s, the largest impediment to establishing a stringent compliance system was the absence of sufficient belief that oil pollution was a problem regarding which an international bargain needed to be struck. It was not a lack of knowledge of oil pollution's sources or impacts: the primary impacts of oil that drive public concern—dead seabirds and oiled beaches—have always been highly visible. Nor did nations lack a knowledge of or a capacity to adopt techniques and technologies that could immediately and effectively have eliminated the problem. States did not find it difficult to make agreements; indeed, states have regularly drafted agreements to control oil pollution since the 1920s. Rather, until the 1970s the major obstacle to the strong international action needed to eliminate coastal oil pollution was a lack of adequate concern—a lack of concern across the majority of states and a lack of sufficiently deep concern in powerful states to lead them to propose strong action and to pressure reluctant states to lend their support.

Three obstacles initially hindered the requisite development of international concern.[126] As these obstacles have been overcome, increasingly stringent regulations have been adopted. One obstacle was the fact that the level of concern among domestic publics was low. In the United Kingdom,

125. Regional Marine Pollution Emergency Response Centre for the Mediterranean Sea, "Review of the Current Situation concerning Reception Facilities for Ship-generated Wastes in Mediterranean Ports," REMPEC/WG.3/INF.5, 7 November 1991.

126. The following discussions draw extensively on Pritchard, *Oil Pollution Control;* United Kingdom Ministry of Transport, *Report of the Committee;* and M'Gonigle and Zacher, *Pollution, Politics, and International Law.*

from the 1920s to the 1960s bird protection societies pressed the government for regulations to halt coastal pollution. In the United States in both the 1920s and the 1970s environmental groups conducted letter-writing, lobbying, and legal campaigns to raise the salience of the issue. However in most states, whether developed or developing, oil pollution was not a public concern that warranted political or financial attention.

Nascent government support for strong measures has often been blocked by domestic oil and shipping concerns. For example, the French and Danish reported serious coastal pollution as early as the 1950s but opposed the 1973 SBT requirements due to their shipping industries' concerns over costs and their shipbuilding industries' concerns that tanker buyers would defer new orders.[127] By the 1950s enough states faced domestic pollution concerns and pressures from activist countries that they were able to require some, albeit low-cost, initial steps to be taken even over the resistance of the oil and shipping interests. However, when states began to make efforts to impose real costs on the oil industry in the 1962 amendments to OILPOL, the affected industries quickly showed their ability to ignore existing regulations and demand less costly rules.

Effective international regulation has also been delayed by difficulties in forming coalitions of nations willing to support strong international action. Through policy compromise and diplomatic pressure, the United Kingdom in 1954 and 1962 and the United States in 1973 and 1978 did build coalitions of nations that were willing to support international controls. Only in the latter two cases, however, was concern deep enough in the activist state, and sufficiently widespread in other countries, that nations agreed to adopt effective international rules. Under environmentalist pressures at home, the United States threatened strong unilateral action if other states did not agree to international regulation. Given the standing of the United States in the international realm, this strategy allowed it to compromise its initial strong positions while still pushing many states well beyond what they would have legislated on their

127. See United Nations Secretariat, *Pollution of the Sea by Oil*; and IMCO, *Pollution of the Sea by Oil* (London: IMCO, 1964) for state positions regarding oil pollution during the 1950s and 1960s. See also M'Gonigle and Zacher, *Pollution, Politics, and International Law*, 86, 90, and 114.

own.[128] Developing countries, many with little domestic experience of or concern over oil pollution, began to support stronger controls because they would feel few of the costs and hoped to establish jurisdictional precedents favorable to their interests in the Law of the Sea negotiations. Developed states lacking strong oil and shipping interests, like Canada, Australia, and New Zealand, also took environmentalist positions and supported the U.S. proposals. By 1973 these changes provided the votes needed for the MARPOL conference to counter the power of maritime states and industry and to pass international regulations that began to require real national and industrial policy responses.[129] While the shift did not occur overnight, the political bargain was no longer weighted exclusively in favor of shipping interests. If prior to the 1970s industry had always dictated oil pollution policy, thereafter it had to negotiate it.

In this chapter I have described the history and contextual factors that led to OILPOL and MARPOL and to the various changes to their compliance systems. In this chapter I have delineated the political context within which these changes were made and the positions of the major actors involved. I have also provided details regarding the technical, economic, and legal aspects of the issue and have described the major actors and incentives at work in oil pollution control. Table 3.4 summarizes major events in this history. Using the information in this chapter as a foundation, in subsequent chapters I analyze whether any of the compliance system changes outlined here have achieved their intended result and, if so, what traits account for their success.

128. In "Environmental Hegemony" Grolin argues that U.S. hegemonic power is sufficient to explain the policies negotiated under IMO auspices as well as the subsequent behaviors.

129. See M'Gonigle and Zacher, *Pollution, Politics, and International Law,* chapters 4 and 5, for an extended discussion of the changes in political bargaining at IMCO between 1954 and 1978.

Table 3.4
International regulation of intentional oil pollution (summary of major events, 1950–1990)

	1950	1960	1970	1980	1990
International Rules					
Adopted	OILPOL (1954)	1962 amendments	1969 amendments MARPOL (1973)	Protocol (1978) 1984 amendments	1987 amendments 1990 amendments
Entry into force		OILPOL (1958)	1962 amendments (1967) 1969 amendments (1978)	MARPOL (1983) 1984 amendments (1985)	1987 amendments (1988) 1990 amendments (1991)
Pollution Major accidents			*Torrey Canyon* (1967) *Argo Merchant* (1976)	*Amoco Cadiz* (1978)	*Exxon Valdez* (1990)
Estimated intentional tanker discharges		Moss 1963 0.45 mta	NAS 1971 1.08 mta	NAS 1980 0.70 mta	NAS 1990 0.16 mta
Crude oil traded	158 mta (1954)	241 mta (1958) 552 mta (1965)	995 mta (1970) 1263 mta (1975)	1320 mta (1980) 871 mta (1985)	1097 mta (1990)
Tanker design and operation		LOT developed (c1962)	COW and SBT developed (c1968)	SBT required (1980) SBT and COW required (1982)	

mta = million tons per annum.

II
Changing Government Behavior

If environmental treaties are to have any beneficial impact on the global environment, they must at a minimum induce governments to change their behavior. Treaties can seek to get governments to reduce those governmental activities that directly harm the environment, to enforce treaty provisions against private activity that harms the environment, or to report on compliance, enforcement, and other treaty-related activities. This section of the book examines efforts made under the International Convention for the Prevention of Pollution of the Sea by Oil (OILPOL) and the International Convention for the Prevention of Pollution from Ships (MARPOL) to improve government reporting on reception facilities and enforcement, to increase government enforcement of treaty regulations against tanker operators and tanker owners, and to get governments to ensure that ports have reception facilities for oil wastes generated by ships.

This section's first chapter examines government reporting and the various systems implemented under OILPOL and MARPOL and under the Memorandum of Understanding (MOU) on Port State Control to elicit reporting. It compares surveys, annual reports, and daily computerized reporting systems used in the OILPOL/MARPOL context and identifies the sources of the successes and failures of these systems to generate comprehensive, high-quality reports. The second chapter of the section details various efforts to improve enforcement by altering the treaties' jurisdictional rules, rights, and obligations and adding equipment standards to the primary rule system. The third chapter examines initial requirements for reception facilities and then considers the efficacy of improvements made to require their installation by specified dates independent of the date of the treaty's entry into force. The section's conclusion (included in chapter 6) identifies common themes regarding the sources of changes in governmental behavior in all three of these areas and the role that various features of compliance systems have played in improving reporting, enforcement, and compliance.

4

Inducing Governments to Report: Making Reporting Easy and Worthwhile

Reporting on compliance, enforcement, and other activities related to environmental treaties is often described as essential to treaty success. Political scientists call for reporting as a means of increasing transparency.[1] Analysts even argue for establishing a climate change reporting system before the convention enters into force to improve compliance.[2] Almost all environmental treaties rely on self-reporting as central to evaluating and improving compliance and call on a secretariat to analyze and review the data.[3] Besides allowing for an evaluation of treaty performance, if reports are made public domestic and international reactions to poor implementation can pressure governments and private actors to improve implementation, compliance, and enforcement. Every international agreement addressing intentional oil pollution has required some form of reporting. The secretariat of the International Maritime

1. See, for example, Abram Chayes and Antonia Chayes, "Compliance without Enforcement: State Behavior under Regulatory Treaties," *Negotiation Journal* 7 (July 1991); Peter H. Sand, "International Cooperation: The Environmental Experience," in Jessica Tuchman Mathews, ed., *Preserving the Global Environment: The Challenge of Shared Leadership* (New York: W. W. Norton and Co., 1991); and Oran Young, "The Effectiveness of International Institutions: Hard Cases and Critical Variables," in James N. Rosenau and Ernst-Otto Czempiel, eds., *Governance without Government: Change and Order in World Politics* (New York: Cambridge University Press, 1991).

2. Abram Chayes and Eugene B. Skolnikoff, "A Prompt Start: Implementing the Framework Convention on Climate Change," a report from the Bellagio Conference on Institutional Aspects of International Cooperation on Climate Change (Cambridge, MA, 28–30 January 1992).

3. Simon Lyster, *International Wildlife Law: An Analysis of International Treaties Concerned with the Conservation of Wildlife* (Cambridge, England: Grotius Publications, 1985), 268.

Organization (IMO) and analysts alike believe that reporting provides information that is "vital to the effective functioning" and proper evaluation of these treaties.[4]

Yet requiring and recognizing the importance of reporting is not the same as ensuring that it is done. Secretariats and environmental advocates regularly criticize countries that fail to provide regular and accurate reports.[5] A U.S. General Accounting Office analysis found that failures to report and inadequate reports were problems common to all the environmental treaties they evaluated.[6] If reporting is essential to achieving treaty aims, how do we make sure that it is done? This chapter evaluates five reporting requirements and systems related to the requirement of reception facilities and the enforcement of this requirement. Comparisons across these systems demonstrate that the incentives to report and the level of countries' development influence all reporting systems, that noncompliance cannot be inferred from nonreporting, and that reporting systems succeed by making data input easy and data output useful to the report providers.

Reporting on Available Reception Facilities

Originally OILPOL required governments to ensure that certain ports had reception facilities adequate to receive the waste oils generated by

4. Clifton Curtis, "Statement," in U.S. House of Representatives, Committee on Merchant Marine and Fisheries, *Hearings on MARPOL Protocol–H.R. 6665: May 8, 1980* (96th Congress, 2nd session) (Washington, DC: GPO, 1980), 605. See also IMCO, *Pollution of the Sea by Oil* (London: IMCO, 1964), 4; George C. Kasoulides, "The Port State Control," paper presented at the Pacem in Maribus XVII Conference, Rotterdam, The Netherlands, August 1990, 6; and MEPC 19/13/7 (6 September 1983).

5. See, for example, Gerard Peet, *Operational Discharges from Ships: An Evaluation of the Application of the Discharge Provisions of the MARPOL Convention by Its Contracting Parties* (Amsterdam: AIDEnvironment, 15 January 1992); Mark C. Trexler, "The Convention on International Trade in Endangered Species of Wild Fauna and Flora: Political or Conservation Success?" unpublished Ph.D. thesis, University of California—Berkeley, Berkeley, CA, December 1989, 56; and Hilary F. French, *After the Earth Summit: The Future of Environmental Governance,* Worldwatch Paper 107 (Washington, DC: Worldwatch Institute, 1992).

6. United States, General Accounting Office, *International Environment: International Agreements Are Not Well Monitored,* GAO/RCED-92–43 (Washington, DC: GPO, 1992).

ships. The convention required governments to notify the treaty secretariat periodically regarding "whether adequate reception facilities have been installed."[7] Resolution 8 of the 1954 conference called on the United Nations "to undertake the collection, analysis and dissemination of . . . information about port facilities for the reception of oily residues."[8] In what became the model for reporting on reception facilities, the U.N. Secretary-General sent a questionnaire to sixty-five states on several oil pollution issues, including the availability of reception facilities in each main port. The survey did not explicitly inquire as to facility adequacy, which in any event the 1954 convention had failed to define. Forty-two states responded in various levels of detail. The United Nations published the responses in 1956.[9]

The 1962 conference deleted the requirement of periodic reporting, but passed a nonbinding resolution that the newly established Intergovernmental Maritime Consultative Organization (IMCO) should obtain and publish information "annually on the progress being made in providing [tanker reception] facilities."[10] Despite these changes, reporting continued to be done through intermittent surveys rather than annual reports. A 1963 IMCO questionnaire asked states to distinguish between oil-loading and other reception facilities, but once again failed to request information regarding adequacy. Of sixty-nine recipients, thirty-two states responded to the questionnaire. IMCO also published the results to inform "shipowners and mariners" of available facilities and thereby increase their usage.[11]

This survey and publication approach to collecting information on available reception facilities has continued through the present. The 1972 survey of eighty countries had received twenty-four responses by the time of the first publication of its results in 1973. Data from three

7. Article VIII, *International Convention for the Prevention of Pollution of the Sea by Oil,* 12 May 1954, 12 U.S.T. 2989, T.I.A.S. no. 4900, 327 U.N.T.S. 3, reprinted in 1 I.P.E. 332, hereinafter cited as *OILPOL 54.*

8. United Nations Secretariat, *Pollution of the Sea by Oil* (New York: United Nations, 1956), 1.

9. United Nations Secretariat, *Pollution of the Sea by Oil.*

10. Resolution 6, IMCO, *Resolutions Adopted by the International Conference on Prevention of Pollution of the Sea by Oil, 1962* (London: IMCO, 1962).

11. IMCO, *Pollution of the Sea by Oil,* 1.

Table 4.1
Reporting rates on availability of reception facilities

	1956	1964	1973/76	1980/84	1985/90
Developed countries (OECD)					
Reports	20	20	19	18	18
Recipients	25	25	25	25	25
Reporting rate (%)	80	80	76	72	72
Developing countries					
Reports	22	14	8	22	19
Recipients	40	44	55	88	93
Reporting rate (%)	55	32	15	25	20
All countries					
Reports	42	34	27	40	37
Recipients	65	69	80	113	118
Reporting rate (%)	65	49	34	35	31

Sources: United Nations Secretariat, *Pollution of the Sea by Oil* (New York: United Nations, 1956); IMCO, *Pollution of the Sea by Oil* (London: IMCO, 1964); IMO, *Facilities in Ports for the Reception of Oil Residues* (London: IMO, 1973, 1976, 1980, 1984); and MEPC Circ. 234 (13 August 1990).

new countries (and updated data from four others) were received and published in 1976. Once again, IMCO published the "information suitably arranged for ships using the facilities."[12] MARPOL reinserted the 1954 requirement regarding self-reporting on available reception facilities. IMCO conducted another survey between 1976 and 1978, publishing the results in 1980 and 1984. Since 1985, surveys on available facilities have been regularly conducted via circulars.[13] Responses since 1985 have been more regularly updated and made more useful to ship operators by including the phone numbers of firms providing reception facilities. The responses have been photocopied, however, rather than published in the glossy pamphlet format of earlier surveys. Table 4.1 shows the number of countries receiving and responding to each of the questionnaires.

The table shows some thirty to forty responses to each of these surveys, even as the number of survey recipients has grown along with the growth

12. IMCO, *Facilities in Ports for the Reception of Oil Residues: Results of an Enquiry Made in 1972* (London: IMCO, 1973), 4.

13. IMCO circulated MEPC/Circ. 38 (16 December 1976) and subsequently MEPC/Circ. 62 (27 January 1978) to request information from governments on reception facilities available in their countries. This information was published in 1980 with a supplement published in 1983.

in the numbers of OILPOL and MARPOL signatories and IMO members. A consistent 70 percent or more of developed states that have received the surveys have responded to them. In contrast, never have more than twenty-five developing states responded, reflecting a significantly lower fraction of recipients in all years. Although it is impossible to verify the accuracy of the data provided, the level of detail and comparability of data across countries has improved considerably between 1956 and today. This improvement has been largely in response to the increasing detail and refinement of the survey questions themselves. Indeed, countries generally have answered the questions asked, and the absence of a question in any of the five surveys on facility adequacy has resulted, not surprisingly, in an absence of any information that indicates adequacy or inadequacy. Since the 1956 and 1963 responses were to questions that were part of more general surveys on oil pollution, in both those years the responses included responses from several countries that indicated they had no reception facilities. Such reporting of no activity provides more information than the absence of a report. Starting in 1973, IMCO has used stand-alone reception facility surveys, and only countries with reception facilities have responded. While the early survey data show countries reporting even while they were out of compliance with the primary rule requiring reception facilities in main ports, more than one-third of countries that reported having facilities in one survey have failed to respond to subsequent surveys, which shows that even states in compliance often fail to report.

Reporting on Inadequate Reception Facilities

Since complying with discharge standards required tankers to retain slops on board and discharge them into reception facilities, oil and shipping interests were very concerned that the use of facilities would involve long and costly delays at the end of each voyage. To address this concern, treaty clauses have consistently required reception facilities to be "adequate to avoid causing undue delay."[14] Although reporting requirements have usually referred to the reporting of information on availablity of

14. Article VIII, *OILPOL 54/62;* Annex I, Regulation 12(1) *International Convention for the Prevention of Pollution from Ships,* 2 November 1973, reprinted

adequate facilities, as just noted, the reporting system has never collected such information.

When negotiators in 1962 deleted the original self-reporting requirement from the 1954 convention, they replaced it with one requiring states to report on other countries' ports with inadequate reception facilities.[15] The United States had proposed this requirement in the hope that it would shame nations into providing facilities by establishing a system for tanker captains, through their governments, to inform IMCO and other governments of absent or inadequate facilities in noncompliant nations.[16] When MARPOL was negotiated in 1973, it included a very similar requirement. However, until 1978 no system was established to collect such information, and IMCO received no reports during the 1960s or early 1970s. Then, hoping to improve this situation, the Marine Environment Protection Committee (MEPC) developed a standardized format for reporting inadequate reception facilities and requested that countries report.[17] A ship's captain could notify the ship's own flag state of inadequate facilities encountered in the ports of other states. Flag state governments were to forward such complaints to the states of the ports concerned and to IMCO.

However, by September 1983 the IMO Secretariat had received only one report on allegedly inadequate reception facilities, and that "through the shipping company and not by a Government."[18] The United Kingdom provided the standard format to its captains, but the reporting

---

in 12 I.L.M. 1319 (1973), 2 I.P.E. 552; and *Protocol of 1978 Relating to the International Convention for the Prevention of Pollution from Ships,* 17 February 1978, reprinted in 17 I.L.M. 1546 (1978), 19 I.P.E. 9451, together hereinafter cited as *MARPOL 73/78.*

15. Article VIII, *OILPOL 54/62.*

16. United States Senate, "Message from the President," *Amendments of the International Convention for the Prevention of Pollution of the Sea by Oil, 1954, March 25, 1963* (88th Congress, 1st session) (Washington, DC: GPO, 1963), 19–20.

17. MEPC/Circ. 60 (16 January 1978).

18. MEPC 19/5 (9 September 1983). See also Y. Sasamura, "Summary of the Requirements of MARPOL 73/78, with Particular Reference to Reception Facilities in Ports," in IMO/UNDP, *Proceedings of IMO/UNDP International Seminar on Reception Facilities for Wastes* (London: IMO/UNDP, 1984), 10–11; and MEPC X/5 (23 October 1978).

procedure was not "used to any great extent either because most reception facilities are adequate or what is more likely because shipping companies are reluctant to risk losing the goodwill of a harbor authority by making an adverse report on the reception facilities."[19]

In 1984 IMO issued a second request for information on inadequate facilities, and this time it received responses from twenty-five countries.[20] In the same year, IMO established an annual report format for all MARPOL-related data that included two stanardized sections, one for reports on inadequate reception facilities referred to other states and one for action taken on such referrals from other states. In the seven years since the new format was established, only four of seventy-five reports have contained any information on inadequate reception facilities.[21]

With the exception of the 1984 survey, in all instances it has been the shipping industry that has supplied IMO with information on inadequate reception facilities. In 1983, 1985, and 1990, the International Chamber of Shipping (ICS) surveyed ship masters and summarized captains' complaints regarding ports where reception facilities were absent, had limited capacity, were costly to use, or required long delays.[22] In contrast to surveys on available facilities, these surveys showed the problems with the facilities in many countries and also identified specific ports that had no facilities at all. The ICS success stemmed from providing a system that protected captains and shipping companies from sanctioning by the harbor authorities mentioned: ICS published the complaints anonymously, thereby receiving considerably more responses than governments did.

On an ad hoc basis, several organizations including IMO have sponsored studies to evaluate the provision of reception facilities, including

19. P. Hambling, "Summary of the Approach Taken in the United Kingdom to the Implementation of MARPOL 73/78 Reception Facility Requirements," in IMO/UNDP, *Proceedings*, 172. See also MEPC 22/8 (1 August 1985), in which the United Kingdom reported four allegedly inadequate reception facilities.

20. IMO made the request in MEPC/Circ. 117 (23 December 1984). Responses can be seen in MEPC 20/Inf. 11 (6 August 1984); MEPC 20/19 (24 September 1984); and MEPC/Circ. 135 (18 March 1984).

21. These data come from the author's survey of all reports received by IMO between 1984 and 1991, contained in various MEPC documents.

22. See the results from the International Chamber of Shipping's 1983, 1985, and 1990 questionnaires reported in MEPC 19/5/2 (21 October 1983); MEPC 22/8/2 (8 October 1985); and MEPC 30/Inf. 30 (15 October 1990).

the absence and inadequacy of facilities.[23] The nongovernmental organization representing independent tanker owners (INTERTANKO) has also "consistently monitored and registered" the lack or inadequacy of reception facilities in many ports.[24] Although IMO recognizes the problems with reception facilities, it has failed to produce a successful compliance information system to address it.

Reporting on Enforcement to the IMCO/IMO Secretariat

Besides reporting on reception facilities, the 1954 convention required that flag states report to IMCO on actions taken on alleged violations referred to it by coastal states. All states were also to provide copies of any reports produced that dealt with treaty compliance and enforcement.[25] As with reporting on reception facilities, reporting on treaty enforcement was initially in response to a 1961 survey by IMCO. Responses from twelve of the thirteen members delineated numerous enforcement difficulties and provided evidence that some states had referred alleged violations to flag states for prosecution.[26] However, none of the flag states receiving such referrals had provided the required reports on actions taken to IMCO.

At the 1962 conference (and again in 1967), hoping to prompt more responsiveness by flag states, the British and French proposed that coastal states provide copies of their referrals to IMCO, but the proposal lacked sufficient support for adoption.[27] The 1962 conference's Resolution 15

23. The Commission of the European Communities and the Italian government sponsored a study on Mediterranean reception facilities, and the Regional Organization for the Protection of the Marine Environment undertook studies on Persian Gulf reception facilities, see IMO/UNDP, *Proceedings;* and P.G. Sadler and J. King, "Study on Mechanisms for the Financing of Facilities in Ports for the Reception of Wastes from Ships," MEPC 30/Inf. 32 (12 October 1990) (Cardiff, Wales: University College of Wales, Cardiff, 1990).

24. See MEPC 27/5 (17 January 1989); and MEPC 27/5/4 (15 February 1989).

25. Article X(2) and Article XII, *OILPOL 54.*

26. OP/CONF/2 (1 September 1961).

27. R. Michael M'Gonigle and Mark W. Zacher, *Pollution, Politics, and International Law: Tankers at Sea* (Berkeley: University of California Press, 1979), 336.

did call on the IMCO Secretariat to produce reports based on national data regarding enforcement, reception facilities, and other treaty-related activities.[28] The 1963 questionnaire discussed above also requested information on enforcement, but through a vaguely worded question regarding "what steps have been taken (other than by laws, regulations, etc.) to prevent oil pollution from taking place."[29] Not surprisingly, only six of the thirty-two responding states even mention enforcement efforts, and these did so only in general terms. Until the 1970s, enforcement reporting was essentially nonexistent. The United Kingdom and Australia had furnished some reports on violations, prosecutions, and penalties in the 1960s, but IMCO had not published them.[30]

The 1973 MARPOL convention retained earlier reporting requirements, but also required that parties provide "an annual statistical report, in a form standardized by the Organization, of penalties actually imposed for infringement of the present Convention."[31] This was the first rule to formally require reports on a fixed schedule.[32] In the wake of the MARPOL Conference, concern over the absence of enforcement reporting grew. In 1973 IMCO specifically requested compliance and enforcement data from states.[33] IMO and its newly formed MEPC also began passing numerous recommendations reminding states of their reporting obligations under OILPOL and MARPOL and urging, though not requiring, them to provide IMCO with copies of referrals to flag stages so they could be "circulated and reviewed regularly," the old French and

28. IMCO, *Resolutions Adopted, 1962.*

29. IMCO, *Pollution of the Sea by Oil,* 7.

30. William T. Burke, Richard Legatski, and William W. Woodhead note their existence in *National and International Law Enforcement in the Ocean* (Seattle: University of Washington Press, 1975), 157, note 253.

31. Article 11(f), *MARPOL 73/78.*

32. The request for annual reporting on inadequate reception facilities was a resolution of the 1962 conference, not an amendment of the 1954 convention's reporting requirements.

33. MPC/Circ. 62 (26 June 1973) cited in Paul Stephen Dempsey, "Compliance and Enforcement in International Law: Oil Pollution of the Marine Environment by Ocean Vessels," *Northwestern Journal of International Law and Business* 6 (1984), 483.

Table 4.2
Reporting rates on OILPOL and MARPOL enforcement

	1975	1976	1977	1978	1979	1980
Developed countries (OECD)						
Reports	3	4	6	7	10	9
Parties[a]	23	23	23	23	24	24
Reporting rate (%)	13	17	26	30	41	38
Developing countries						
Reports	0	2	3	3	4	7
Parties	31	35	36	37	38	41
Reporting rate (%)	0	6	8	8	10	17
All countries						
Reports	3	6	9	10	14	16
Parties	54	58	59	60	62	65
Reporting rate (%)	6	10	15	16	22	24

Sources: Paul Stephen Dempsey, "Compliance and Enforcement in International Law: Oil Pollution of the Marine Environment by Ocean Vessels," *Northwestern Journal of International Law and Business* 6 (1984), 485; and Gerard Peet, *Operational Discharges from Ships: An Evaluation of the Application of the Discharge Provisions of the MARPOL Convention by Its Contracting Parties* (Amsterdam: AIDEnvironment, 15 January 1992), annexes 4, 5, and 13.

[a]Number of nations party to *either* OILPOL or MARPOL. The number of OILPOL parties declines after 1987 due to denunciations by Australia, Germany, the Netherlands, and Bulgaria.

British proposal.[34] MEPC developed a form for reporting penalties imposed under MARPOL in 1974.[35]

Coupled with the increased attention to oil pollution generated by MARPOL and the semiannual MEPC meetings, these requests elicited the first annual reports in 1975.[36] Greece provided reports on referrals it had made to other flag states and actions it had taken in response to referrals from coastal states, and it was joined by Ireland and the United

34. M'Gonigle and Zacher, *Pollution, Politics, and International Law,* 337. For example, Resolution A.392(X) (14 November 1977) and Resolution A.542(13) (17 November 1983) called on governments to report on enforcement more regularly, including notifying IMCO of all referrals to flag states.

35. MEPC II/5 (25 July 1974).

36. See Burke, Legatski, and Woodhead, *National and International Law Enforcement,* 157; and Dempsey, "Compliance and Enforcement."

Table 4.2
(continued)

1981	1982	1983	1984	1985	1986	1987	1988	1989	1990
4	na	9	10	15	13	12	17	15	15
24	24	24	23	23	23	23	22	21	22
17	na	38	43	65	56	52	77	71	68
8	na	4	7	15	16	12	15	17	19
43	44	47	48	49	49	49	49	49	47
19	na	9	15	30	32	24	30	34	40
12	na	13	17	30	29	24	32	32	34
67	68	71	71	72	72	72	71	70	69
17	na	18	23	41	40	33	45	45	49

Kingdom in reporting actions taken against violations within territorial waters. Table 4.2 shows that enforcement reporting increased gradually after these initial reports. The IMCO Assembly reiterated the need for national reports on prosecutions and called on the secretariat to compile these.[37] However, by 1981 fewer than one-third of the parties to OILPOL had filed annual reports, most having "ignored the requirement altogether."[38] Since IMCO had not specified the form or content of these reports, many reports provided "little meaningful information." Both quality and content varied immensely across countries, and even across time for a single country, making detection of trends or other assessments difficult. Even the time periods covered by reports were unspecified or inconsistent.[39]

Concern about enforcement of IMCO conventions in general and MARPOL in particular increased in the early 1980s just prior to MARPOL's entry into force. Given the major revisions of MARPOL, IMCO sought to shift its focus from legislation to implementation.[40] Reporting

37. Resolution A.392(X) (14 November 1977).
38. Dempsey, "Compliance and Enforcement," 484.
39. Dempsey, "Compliance and Enforcement," 463.
40. Resolution A.500(XII) (20 November 1981).

was one major concern. Although MEPC had been tasked to develop a standardized reporting format by the 1973 convention, it completed the task only in 1985.[41] The specification of report content and format has both increased report response rates and vastly improved report comparability. The pre-1985 average of eleven countries reporting per year has almost tripled to an average of thirty since 1985. Nevertheless, many countries still fail to report, and even states that do report often fail to complete all the sections of the forms. A 1992 study found that six parties had submitted reports every year, and three more had submitted them for every year minus one, but that more than thirty had never submitted a report, while the remainder had submitted "often incomplete reports for one or a few years only."[42] Thus, reporting problems have continued to plague IMO.

In contrast to the consolidation and publication of reports on reception facilities, IMO's analysis of reports on enforcement has been infrequent and of poor quality. IMO did not produce its first analysis of enforcement reports until twenty-nine years after the initial requirement for reports was established and eight years after it had begun receiving reports. The 1983 analysis aggregated data from 1977 through 1981 in a single two-page report with an attached four-page summary chart of violations reported, investigated, and prosecuted.[43] The Secretariat claimed that "it has not been possible to discern particular trends." The Secretariat's next analysis was five years later and evaluated a single year's data only. Again it was brief and unenlightening.[44] In all other years, the Secretariat has merely copied and distributed national reports at MEPC meetings, some-

41. The new three-part format was first circulated by MEPC/Circ. 138 (15 May 1985) and recirculated by MEPC/Circ. 228 (29 March 1990). Part II, an Annual Enforcement Report, contained sections on violations referred to flag states, actions taken on such referrals from other states, inadequate reception facilities referred to flag states, and actions taken on referrals of inadequate reception facilities. Part III was an Annual Assessment Report, sections 1 and 2 of which included statistics and details on port state control inspections, deficiencies and detentions, section 3 of which addressed penalties imposed for violations.

42. Peet, *Operational Discharges,* 5–6.

43. MEPC 19/13/7 (6 September 1983).

44. MEPC 26/14 (27 July 1988).

times provoking discussion but not significant efforts to induce greater reporting.[45]

All significant analysis of the enforcement data that has occurred has been undertaken by outside sources. A law journal article published in 1984 and using the same data as the 1983 IMO study conducted a thorough analysis of trends in enforcement and compliance as well as identifying problems with the reporting system itself.[46] The 1992 study mentioned above, was conducted by the Friends of the Earth representative to IMO, and provides an in-depth and systematic analysis of the data in the standardized reports available to IMO.[47]

Reporting on Enforcement to the Memorandum of Understanding Secretariat

The purpose of the Memorandum of Understanding (MOU) on Port State Control was to increase the enforcement efforts of all IMO conventions, including MARPOL, by the fourteen European member states. The agreement requires members to inspect 25 percent of the foreign ships entering their ports and provide data on those inspections to a centralized computer data base, "preferably . . . by means of direct, computerized input on a daily basis."[48] Through telex and modem links rather than annual reports, inspectors report daily to a central computer in France on inspections conducted and violations detected.

45. Certainly nothing like the International Labor Organization's practice of blacklisting nonreporters has ever been attempted; see E. A. Landy, *The Effectiveness of International Supervision: Thirty Years of I.L.O. Experience* (London: Stevens and Son, 1966).

46. Dempsey, "Compliance and Enforcement."

47. Unfortunately the analysis covers only the period since MARPOL entered into force in 1983 (Peet, *Operational Discharges*).

48. Section 4 and Annex 4(2), *Memorandum of Understanding on Port State Control in Implementing Agreements on Maritime Safety and Protection of the Marine Environment*, 26 January 1982, reprinted in 21 I.L.M. 1 (1982), I.P.E. II/A/26-01-82, hereinafter cited as *MOU*. For each ship inspected a telex message provides information on the issuing country; the name, type, flag, call sign, tonnage, and year of build; the date and place of inspection; the nature of deficiencies; and any action taken.

The MOU Secretariat uses this already digital data in two ways. First, national inspectors can use modems to query the data base for information regarding specific tankers entering their ports. The date and results of the most recent inspection by any member country or the absence of any data provides inspectors with valuable information that helps them decide which tankers to inspect. Authorities can thereby deploy scarce inspection resources more effectively and efficiently. Second, the MOU Secretariat aggregates and extensively analyzes the individual inspections for its annual reports. These reports provide detailed summaries from the current and previous two years of inspections, deficiencies found, and penalties imposed. Data are not broken down by country, although the enforcement activities of individual countries are analyzed during annual meetings of the Port State Control Committee, the main administrative arm of the MOU.

For every year since the MOU took effect in 1982, all fourteen members have provided data on many of their inspections.[49] Even the first year over 8,800 inspections of 6,300 different ships were reported, and those numbers have grown steadily every year since. Since an annex to the MOU agreement itself specifies a simple format for the data, the data is impressively consistent, allowing for comparisons of data from year to year and country to country. Of course, what evidence exists suggests that data quality is improving, more often including all inspections conducted and accurately reflecting inspection results. Data quality is attested to by the fact that most states have consistently reported, even though the total number of inspections indicates that very few MOU states have ever met the required 25 percent inspection rate. The use of country-level statistics only in diplomatic sessions, the aggregation of inspections into annual rates by the secretariat rather than by national governments, and the value of the data base in making decisions regarding future inspections eliminate incentives to falsely overstate inspection rates.

49. See the Secretariat of the Memorandum of Understanding on Port State Control's *Annual Report* (The Hague: The Netherlands Government Printing Office), hereinafter cited as Secretariat, *Annual Report,* for all years since 1982/83.

Analysis

In essence, five quite different reporting systems have been detailed here. Using surveys, the system for reporting reception facility availability elicits relatively high-quality reports from many nations, including both developed and developing countries. Reporting on inadequate reception facilities has been essentially nonexistent by governments, and IMO data have come almost exclusively from independent surveys and ad hoc studies of specific regions. Initially IMO received a few inconsistent and low-quality enforcement reports despite the absence of a systematized reporting system. Since a standardized format was developed in 1985 and institutional attention to such reports increased, both the number and the quality of national reports have increased, although there have consistently been fewer than twenty reports per year. More than any of these systems, however, the MOU's system for daily enforcement reporting has elicited regular, high-quality reporting by all the states involved.

How do we sort out the factors responsible for the quite different levels of reporting under each of these systems? Comparisons within and across the systems show how reporting depends on differences in the capacity of states to report, the willingness of states to divulge information about the activity they are reporting on, and the qualities of the reporting systems themselves.

A country's level of development dramatically influences the likelihood that it will report. As both tables 4.1 and 4.2 show, developed states are far more likely both numerically and proportionally to report than are developing states. On both available reception facilities and enforcement, developed states have reported to IMO at rates two to three times those of developing states. This consistent disparity supports evidence from treaties on other issues that developing states often have inadequate financial and administrative capacities and domestic concern to report.[50] The fact that such states often also lack the resources and concern to comply with substantive primary rules (by providing reception facilities or enforcing) reinforces the absence of reporting. The fact that many

---

50. Abram Chayes and Antonia Handler Chayes, "On Compliance," *International Organization* 47 (Spring 1993), 200; and General Accounting Office, *International Environment*.

developed states fail to report, however, documents that capacity is a necessary but by no means sufficient condition for reporting.

Reporting also varies with incentives to report. The contrast between the absence of governmental reporting on inadequate reception facilities and the frequent reporting by shipping interests shows how incentives and report frequency vary according to what is being reported on and who is reporting. Most diplomats have few incentives to confront other diplomats regarding the inadequacy of reception facilities since few countries have impeccable records in this regard and since such inadequacy causes little immediate harm to other governments. In contrast, oil and shipping interests are directly harmed by long delays at small and inefficient discharge facilities. Besides, governments have broader goals that make them less likely than industry to rely on unverified reports of inadequacy by tanker captains and to provide those captains with the anonymity they need to protect them from reprisals by harbor masters. Not surprisingly, industry groups have produced the majority of data on inadequate facilities.

To identify the impact of different reporting systems, therefore, requires that we control for the levels of development and the incentives of the actors responsible for reporting. Comparing the responses of the fourteen MOU states to both the MOU and the IMO enforcement reporting systems allows us to look at variations in reporting on the same activity by the same group of states over the same period of time. These were all developed states that were enforcing rigorously. Table 4.3 shows that all fourteen states have reported to MOU every year, while no more than nine states have ever reported to IMO in one year and the average number of reports per year since 1982 has been fewer than seven.[51] Not only does this controlled comparison eliminate other sources of variation, but the differences in the reporting systems readily explain the different reporting rates.

51. Despite being established to improve compliance with the MARPOL agreement, among others, the MOU's computerized information system "does not entirely meet the requirements of the IMO." Only in 1990 did the Secretariat undertake a feasibility study to assess how MOU reporting requirements could be adjusted to enable member states to fulfill IMO and MOU reporting obligations simultaneously, but these adjustments have yet to be made (Secretariat, *Annual Report 1990,* 16; and Secretariat, *Annual Report 1992,* 25).

Table 4.3
Responses by MOU states to self-reporting systems[a]

	1982	1983	1984	1985	1986	1987	1988	1989	1990
Total number of MOU states	14	14	14	14	14	14	14	14	14
Reports on enforcement to MOU	14	14	14	14	14	14	14	14	14
Reports on enforcement to IMO	4	7	7	9	8	6	7	7	6
Reports on reception facilities to IMO		13							13

[a]This chart depicts the number of reports received from the fourteen member states of the Memorandum of Understanding (MOU) on Port State Control. For example, in 1990 all fourteen MOU states filed reports on enforcement with the MOU, six MOU states filed reports on enforcement with IMO, and thirteen MOU states filed reports on reception facilities with IMO.

The MOU reporting system has four characteristics that, taken together, explain the higher response rates of that system. First and most important, the MOU used the data it received to create information that was valuable to those actors providing the data. The MOU system helped the bureaucrats doing the reporting by helping them choose which incoming ships to inspect. Distinguishing ships that had not been recently inspected from those recently inspected and found deficient or those recently inspected and found in good shape allowed inspectors to focus limited enforcement resources on likely violators while avoiding inspections of likely compliers.[52] While such information was incredibly valuable, no state could generate such information alone. The computerized system also allows one inspectorate to inform another instantly and easily that a ship is suspected of having committed a violation, enhancing coastal pollution control. In short, the MOU design makes reporting part

52. The MOU specifically provides that "the Authorities will seek to avoid inspecting ships which have been inspected [and found to have no deficiencies] by any of the other Authorities within the previous six months, unless they have clear grounds for inspection" (Section 3.4, *MOU*).

of an integrated information system that both maritime authorities and the MOU Secretariat can use as a management tool to deploy and evaluate enforcement resources.

Another characteristic of the system is that, while authorities had these incentives to provide the information, the MOU also made data requirements specific and simple and created an easy-to-use computerized data entry system. In contrast, it took IMO until 1985 to develop a standard data format telling countries what information to report. Lacking an obvious form to fill out, most countries failed to develop one on their own.

A third characteristic is that the far greater demands of daily reporting compared to annual reporting paradoxically helped increase reporting: meeting a daily data entry requirement forced national inspectorates to establish standard operating procedures. Whether this involved minor modification of existing management information systems or the establishment of a new system, the MOU system reduced the costs of maintaining national systems by allowing them to piggyback on it.

Finally, the MOU Secretariat has made reporting seem essential to the effectiveness of the agreement by analyzing the data and creating and disseminating summary reports that make use of the enforcement and compliance information that is reported. In contrast, IMO does little more than duplicate and distribute the reports it receives. Frequent postponement of reviews of reports at IMO meetings gives countries a sense that failure to report has no significant effect on treaty operation. IMO's failure to monitor reporting and either sanction or assist countries failing to report reinforces this view. IMO neither publishes a "blacklist" of countries failing to report, nor does it publish a "white list" of countries that have reported.[53]

Whether these system characteristics can produce similar results among developing states will be tested in the near future as the Latin American Port State Control regime establishes a similarly designed reporting system.

A similar comparison of the response rates of these same MOU states to IMO's surveys on available reception facilities shows an alternative to

---

53. The International Labor Organization makes extensive use of such lists to encourage reporting (Landy, *The Effectiveness of International Supervision*).

daily data entry to successfully elicit reporting. Among MOU states—states largely complying with the rules regulating both reception facilities and enforcement—only three reported on enforcement in 1975 through 1976, between five and seven reported in 1977 through 1984, and between six and nine reported in 1985 through 1990. In each of these same periods, thirteen MOU states responded to IMO's surveys on available reception facilities. The fact that these countries report on enforcement to MOU confirms that the variation is due not to *inherent* differences between the incentives to report on facilities and those to report on enforcement, but to the specific differences in incentives *created by* the reporting systems. Of all developed (OECD) countries, eighteen to twenty have responded to each survey, although in every year but one, fifteen or fewer have provided enforcement reports.

Where the MOU's enforcement reporting succeeded by relying on standard operating procedures and technology to ease a very demanding reporting burden, IMO's surveys on reception facilities succeeded because IMO placed a specific report format in bureaucrats' in-boxes that imposed only a slight reporting burden on the reporting entities and did so only infrequently. Reporting on reception facilities required most countries to collect specified information on only five to ten ports and fifteen to twenty-five reception facilities. The surveys succeeded because they required little bureaucratic effort and because answering the specific survey questions did not require a government to have or create a standing informational infrastructure. IMO also collated, published, and disseminated the information in a format that made it accessible to tanker owners and operators, who might therefore be more likely to use reception facilities. In contrast, acquiring accurate annual enforcement data requires the creation of a large management information system to track hundreds or thousands of inspections, prosecutions, etc., involved in enforcement. Until 1985, IMO had not yet specified what information should be collected. Unless the data would fulfill other missions, a government would be unlikely to create such a system merely to produce one report per year for an international organization that made little use of it and imposed no sanctions for not reporting. Surveys would be unlikely to work for enforcement reporting, although they provide an appropriate mechanism for reporting on reception facilities.

Data on both enforcement and reception facilities document that countries often do not report even when they are complying with underlying primary rules. Nonreporting is not equivalent to noncompliance. Of fourteen states that were enforcing rigorously and meeting MARPOL's enforcement requirements, half regularly failed to report that fact to IMO. The United States referred numerous OILPOL violations to flag states between 1969 and 1977 and published data from its computerized oil spill information system throughout the 1970s, but it failed to provide enforcement reports to IMO in any of those years.[54] Canada also made some eighty referrals from 1967 to 1977, but did not report any of them to IMO.[55] The Dutch have kept oil spill statistics since 1969 and began aerial surveillance in 1975, yet reported to IMO on enforcement in only three years prior to 1982.[56] Given Germany's extensive enforcement as reported in the 1961 survey and in more recent years, the absence of German reports to IMO during the late 1960s and early 1970s most likely reflects nonreporting rather than nonenforcement.[57] Belgian national records show inspections of tankers in every year from 1984 to 1988, but Belgium provided enforcement data to IMO only in 1988.[58] The data on reception facilities confirm this finding; several countries

54. From 1969 to 1972 the United States referred seven cases to flag states (Burke, Legatski, and Woodhead, *National and International Law Enforcement*, 202). From 1972 to 1977 the U.S. referred between five and fifty-seven violations each year, yet reported none of these to IMCO, as can be seen by comparing U.S. House of Representatives, Committee on Merchant Marine and Fisheries, *Hearings on Port Safety and Tank Vessel Safety: April 6, 13, 20, and June 15,* (95th Congress, 2nd session) (Washington, DC: GPO, 1978), 506, with Dempsey, "Compliance and Enforcement." On the U.S. oil spill information system, see reports regularly published since 1973 by the United States Coast Guard entitled *Polluting Incidents in and Around U.S. Waters* (Washington, DC: GPO). IMO received only the first of these.

55. M'Gonigle and Zacher, *Pollution, Politics, and International Law,* 334.

56. Compare N. Smit-Kroes, *Harmonisatie Noordzeebeleid: Brief van de Minister van Verkeer en Waterstaat* (Tweede Kamer der Staten-Generaal: 17–408) (The Hague: The Netherlands Government Printing Office, 1988) with Dempsey's "Compliance and Enforcement."

57. See Dempsey, "Compliance and Enforcement," on the absence of German reporting.

58. Compare Greenpeace, *Implementation of the Second North Sea Conference: Belgium* (Greenpeace Paper 11) (Amsterdam, The Netherlands: Greenpeace International, 1989), 82–83, with Peet, *Operational Discharges,* Annex 2.

reported having reception facilities in early surveys but failed to respond to subsequent surveys, despite the fact that it would be unlikely that they had eliminated their facilities.

Showing that compliant states do not always report and that reporting systems can create incentives for them to do so is not equivalent, however, to showing that such systems can clear a higher hurdle by eliciting reporting from those that have not complied with the primary rules. Nevertheless, the evidence supports the conclusion that noncomplying states often do report. The MOU states regularly report to the MOU Secretariat even though they know they will not meet the required 25 percent inspection rate. Industry reports strongly suggest that many governments answer IMO surveys on available reception facilities even though facilities are clearly inadequate. In both cases, noncompliant reporters face no significant pressures for having failed to comply. In contrast, tanker captains have been unwilling to report on the inadequacy of reception facilities they encounter unless they are assured of anonymity to protect them from retaliatory actions by harbor authorities. The condition for eliciting reporting from actors, whether self-reporting on one's own compliance or identifying noncompliance by others, appears to be an implicit or explicit assurance that reporting will not prove more costly than not reporting.

Conclusions

States clearly do not report simply because a treaty requires it. Noncompliance with reporting requirements has been common in all the OILPOL- and MARPOL-related reporting systems. States that fail to report cannot, however, simply be assumed not to be complying with the rules regulating the behavior in question. Nonreporting is not equivalent to noncompliance. Many countries have provided information to the IMO and MOU Secretariats that show their noncompliance with the primary rules; others have failed to provide reports although they were in substantial compliance with those rules.

As documented with regard to other environmental treaties, self-reported information is difficult to elicit.[59] Considerable disincentives to

59. General Accounting Office, *International Environment*.

reporting exist. In self-reporting systems, states or other actors will not report if they believe that revealing their own noncompliance will prove more costly than remaining silent. Even nonstate actors that are reporting on noncompliance by governments, like tanker captains reporting on inadequate reception facilities, may not report unless they can be protected from retaliatory actions by those governments. The difficulty of determining what information is required and then gathering and aggregating that information may be great enough that it discourages, if not prohibits, many governments from responding.

The wide variations in reporting under the various systems described here show that success at eliciting reporting depends on addressing these concerns and providing actors with positive incentives to report. Obviously those actors consciously opposed to a treaty's goals will violate its terms and cannot be expected to report. However, reporting systems can aspire to elicit reports not only from those actors complying with the treaty's primary rules, but also from those failing to comply because of inadvertence, incapacity, or other unintentional causes. MOU produced incentives to report by creating a system of sufficiently current and extensive data that reporting agencies provided information to it because it fed back information that helped them enforce better. IMO's reception facility publications made reception facility usage more likely. In these systems reporting was not post hoc paperwork, but a means of furthering the primary organizational missions of the reporting agencies.

In contrast, IMO's reporting on enforcement and inadequate reception facilities has languished. The systems used for this reporting not only failed to use the data in ways that facilitated the reporting actors' meeting their goals, but also failed to convince them that the reported information fostered treaty effectiveness. At the same time, collecting and providing the information involved considerable cost. Lack of capacity certainly explains why some countries fail to report, but lack of incentives also plays a crucial role. The MOU created a system that induced numerous developed countries to report even while they were failing to do so under the IMO system by making reporting useful and easy and by assuming many of the costs of data storage and analysis. Although we often think of reporting as a process whereby states and nonstate actors provide information to a treaty secretariat, successful reporting systems involve ensuring that the information coming in also goes back out in a form that the original information providers find useful.

Notable due to its absence from these five reporting systems has been the use of shaming and blacklisting to increase reporting. Neither governments nor intergovernmental organizations have sought to shame governments into reporting through publicizing lists of nonreporters.[60] MOU does not publish country-by-country data; my knowledge that all countries have reported came from interviews.[61] In its reception facility publications, IMO lists only countries reporting, not nonresponding recipients. While analysts have noted that "recording, publication, and discussion of enforcement records could put substantial pressure on" countries to enforce, IMO has put little effort into such tasks.[62] Nonreporting itself, although simple to monitor, has never been highlighted by the IMO or MOU Secretariats. Governments have diplomatic and other disincentives to shame other governments. What shaming is done, therefore, is done by nongovernmental actors with the incentives and ability to monitor and criticize governmental noncompliance. A Friends of the Earth International analysis of reporting prompted MEPC discussions that appear likely to lead to regularly listing states that have and have not reported.[63]

Finally, providing clear standardized formats for reports is critical if only so secretariat personnel can analyze the data once it has been received. The absence of standard operating procedures in most countries for generating annual enforcement reports can be overcome either by

60. See, for example, Robert O. Keohane, *After Hegemony: Cooperation and Discord in the World Political Economy* (Princeton, NJ: Princeton University Press, 1984), 105; and Chayes and Chayes, "Compliance without Enforcement," 323.

61. See George Kasoulides, "Paris Memorandum of Understanding: A Regional Regime of Enforcement," *International Journal of Estuarine and Coastal Law* 5 (February 1990); and Secretariat, *Annual Report*, 1982/83–1992. Similarly, under the Mediterranean Action Plan "environmental quality data are aggregated for regions . . . so as not to embarrass individual coastal states"; see Peter M. Haas, "Do Regimes Matter? Epistemic Communities and Mediterranean Pollution Control," *International Organization* 43 (Summer 1989), 383.

62. M'Gonigle and Zacher, *Pollution, Politics, and International Law,* 336.

63. Gerard Peet, personal communication, 1992. Whether this will increase reporting remains to be seen.

establishing and facilitating daily reporting routines, as in the MOU case, or by mailing specific surveys that prompt responses from national bureaucracies. Computer technologies foster the provision of reports to member states in electronic formats that make data base maintenance easier and allow reporting agencies to use the formats for day-to-day tasks and not just to generate annual reports.

Reporting system design can dramatically alter both the number and the quality of the reports nations and other actors provide. Successful reporting systems address existing disincentives to report while providing reporting entities with both simple and specific data requests and positive incentives to fulfill them. Only with such data as the empirical foundation can those interested in learning from past experience identify how effectively a treaty is altering behavior and achieving its goals.

5

Inducing Governments to Enforce: Deterrence-Based versus Coerced Compliance Models

Treaties drafted to reduce oil pollution, like those designed to address many other environmental problems, face a two-level regulatory challenge: to achieve compliance, such treaties must cause governments to adopt, implement, and enforce rules in ways that, in turn, cause private actors to adopt new behaviors.

As noted in chapter 2, several recent treaties have noncompliance response systems that depend on governments to create systems of positive inducements for private actors to adopt new behaviors. Such systems include the financial mechanisms of the Montreal Protocol and the tax or tradable permit schemes proposed with regard to the Framework Convention on Climate Change. In contrast, response systems of the International Convention for the Prevention of Pollution of the Sea by Oil (OILPOL) and the International Convention for the Prevention of Pollution from Ships (MARPOL) have relied more on enforcement to induce compliance, making violation either more difficult or more costly.[1] As oil tanker operators have proven unlikely to change their behavior on their own and governments have proven unwilling to pay them to do so,

---

1. *International Convention for the Prevention of Pollution of the Sea by Oil,* 12 May 1954, 12 U.S.T. 2989, T.I.A.S. no. 4900, 327 U.N.T.S. 3, reprinted in 1 I.P.E. 332, hereinafter cited as *OILPOL 54;* and *International Convention for the Prevention of Pollution from Ships,* 2 November 1973, reprinted in 12 I.L.M. 1319 (1973), 2 I.P.E. 552; and *Protocol of 1978 relating to the International Convention for the Prevention of Pollution from Ships,* 17 February 1978, reprinted in 17 I.L.M. 1546 (1978), 19 I.P.E. 9451, hereinafter cited together as *MARPOL 73/78.*

"the only solution to the deliberate discharge of pollutants is the brute force of strict enforcement."[2]

The two primary rule systems established in OILPOL and MARPOL—discharge standards and equipment standards—rely on quite different models of enforcement. The compliance system for the discharge standards relies primarily on creating credible and potent threats to violation—a deterrence-based model of enforcement. This system is intended to increase compliance by making violation more costly. The compliance system for the equipment standards instead focuses on creating obstacles to violation to induce compliance—a coerced compliance model.[3] This system is intended to increase compliance by making violation more difficult. This chapter compares these two systems as well as examining the various efforts made to improve them. It examines the importance of enforcement in these treaties, both as a form of governmental compliance with treaty enforcement requirements and as the means of eliciting compliance by private actors.

Enforcement of Discharge Standards

In choosing to attack the problem of intentional oil pollution by limiting discharges, the drafters of the 1954 OILPOL in large measure predetermined the type of enforcement system that would be needed to induce compliance. The 1954 convention required tanker operators to ensure that all discharges within "prohibition zones" extending fifty miles from shore not exceed one hundred parts of oil per million parts of water (100 ppm). This definition of violations in terms of the location and oil content of the discharges determined the fundamental structure of the enforce-

2. Albert G. Stirling, "Prevention of Pollution by Oil and Hazardous Materials in Marine Operations," in *Proceedings: Joint Conference on Prevention and Control of Oil Spills, 15–17 December 1969* (New York: American Petroleum Institute, 1969), 48. See also Sir Colin Goad in R. Michael M'Gonigle and Mark W. Zacher, *Pollution, Politics, and International Law: Tankers at Sea*, (Berkeley: University of California Press, 1979), 327; and Resolution A.391(X) (14 November 1977).

3. Albert J. Reiss, Jr., "Consequences of Compliance and Deterrence Models of Law Enforcement for the Exercise of Police Discretion," *Law and Contemporary Problems* 47 (Fall 1984).

ment problem that governments would face and that subsequent amendments would seek to remedy.

The compliance system for these discharge standards was activated only after a violation of the standards had been committed, and it consisted of three phases. First, governments had to detect violations and identify violators. The nature of the oil transportation business—both the large number of tankers and the diversity of possible routes—makes detecting violations of discharge standards an enormous task. The original definition of violation in terms of the oil content of the discharge required reliance on the discharging tanker's own on board measurement equipment, since oil content could not be measured once the discharge stream had entered the water. Although any large, visible slick could be assumed to violate the standard, the establishment of fifty-mile coastal prohibition zones meant that huge areas of ocean had to be patrolled. Even if such a slick was detected, it was difficult to identify the tanker responsible. Under the initial standards there was frequently a gap between detected violations and identified violators.

In the second phase of the compliance system, governments had to prosecute and convict identified violators. However, information that linked a specific tanker with a detected discharge violation often could prove inadequate as legal evidence of that fact. This problem, which is common in any legal system, is compounded internationally: evidence that is adequate in one nation's courts may prove inadequate in another's. In the third phase of the compliance system, governments had to penalize convicted violators. In practice this often proved difficult since a tanker successfully convicted of a discharge violation often left the port of a prosecuting state long before a sentence or fine had been imposed.

The problems with this system for enforcing zonal discharge limits, which had been recognized as early as 1926, were highlighted in the responses to a 1961 survey conducted by the Intergovernmental Maritime Consultative Organization (IMCO).[4] Of the twelve parties to the

4. Charles Hipwood, head of the U.K. Marine Department, has noted, "We know the difficulties of getting evidence within our own three-mile limit. *A fortiori* what are the difficulties going to be in enforcing it when it comes to a matter of 50 to 150 miles?", cited in Sonia Zaide Pritchard, *Oil Pollution Control* (London, England: Croom Helm, 1987), 23.

1954 convention, essentially only Britain and West Germany were detecting and prosecuting violations. Of 705 detected violations, more than 80 percent involved discharges in port or within three-mile territorial seas. Of these, about half produced convictions, mostly in cases of discharges in port rather than at sea. In most cases, the fines levied were quite small.[5] Only 128 (18 percent) of the violations were committed outside territorial waters. The responsible flag states failed to successfully prosecute a single one of these. Not unexpectedly, states complained of several enforcement problems: assessing whether an observed discharge violated the standard of 100 ppm, patrolling large areas of ocean, captains' falsifying the oil record books (ORBs) they were required to maintain, and flag states' failing to prosecute cases referred to them.[6]

Besides the practical enforcement problems, existing international legal norms obstructed effective enforcement. Recognizing that rules adopted in OILPOL might be used as precedents by countries seeking to expand their legal jurisdiction in other issue areas, powerful maritime states ensured that the agreement relied on the existing two-tiered enforcement jurisdiction of customary international law. For violations outside their territorial seas, port and coastal states had the right (but not the obligation) to collect evidence of the violations for referral to the flag states. The flag states had the obligation to investigate such referrals and to prosecute those for which they were "satisfied that sufficient evidence [was] available."[7] For violations within their territorial seas, port states, but not coastal states, had the right to investigate, prosecute, and penalize ships.

Given such practical difficulties in detecting violations and identifying violators and such legal difficulties in prosecuting, convicting, and penalizing violators, it was not surprising that the 1961 survey showed that detection of violations beyond territorial seas and prosecution by flag states was essentially nonexistent, despite the treaty's fifty-mile zones. Since then, various efforts have been made to remedy problems in one or more of the three phases of enforcement.

5. OP/CONF/2 (1 September 1961).

6. Pritchard, *Oil Pollution Control*, 112; and M'Gonigle and Zacher, *Pollution, Politics, and International Law*, 220–221.

7. Article IX and X, *OILPOL 54*.

Improving Detection of Violations and Identification of Violators

Before the compliance system could deter tanker operators from illegal discharges, it had to make detection and identification of violations more likely. Efforts met initial resistance. At the 1962 amendment conference, British and French proposals to increase port states' inspection rights "engendered vehement opposition."[8] States were not yet willing to sacrifice any of their sovereignty to enhance marine protection. By the late 1960s, there were still no oil content meters with which captains, let alone others, could accurately monitor compliance with the 1954 standard of 100 ppm.[9] Increased concern over enforcement and the increased willingness of states to take measures to address that issue became evident in resolutions adopted by the IMCO Assembly in 1968. Resolution A.151 "invited" port states to increase their detection efforts, including undertaking inspections of ships that any other state reported as having violated the convention. The French pressed especially hard for new enforcement measures, proposing a requirement that all states inspect all vessels in their ports and board vessels near their coasts to search for possible discharge violations.[10] Denmark, Finland, Norway, and Sweden signed a new agreement to cooperate on the detection of violations and the enforcement of OILPOL regulations in 1969.[11] These factors, plus U.S. pressures for a more enforceable convention and industry desires to avoid expensive equipment requirements resulted in the 1969 amendments.[12]

The 1969 amendments to OILPOL sought to redefine the discharge standards so they would be more verifiable. They replaced the 100 ppm

8. M'Gonigle and Zacher, *Pollution, Politics, and International Law,* 221; and United States Senate, "Message from the President," *Amendments of the International Convention for the Prevention of Pollution of the Sea by Oil, 1954* (88th Congress, 1st session) (Washington, DC: GPO, 1963), 47.

9. This remained true until at least 1976; see MEPC VI/4 (30 September 1976), 2.

10. M'Gonigle and Zacher, *Pollution, Politics, and International Law,* 224.

11. *Agreement between Denmark, Finland, Norway, and Sweden Concerning Cooperation to Ensure Compliance with the Regulations for Preventing the Pollution of the Sea by Oil, Copenhagen,* 8 December 1967, 620 U.N.T.S. 226, reprinted in Bernd Ruster and Bruno Simma, eds., *International Protection of the Environment: Treaties and Related Documents,* Volume 1 (Dobbs Ferry, NY: Oceana Publications, 1975), 446.

12. United States Senate, "Message from the President," 39.

standard with a "clean ballast" criterion for discharges within prohibition zones and added a new standard for discharges outside these zones, which was defined in liters discharged per mile. The new rules prohibited any discharge within fifty miles of land except one that, "if it were discharged from a stationary tanker into clean calm water on a clear day, would produce no visible traces of oil on the surface of the water."[13] This "clean ballast" provision was hailed as remedying detection problems: sighting a spill now equated, by definition, to detecting a violation. Such a change did not address the fact that most signatories had failed to establish detection programs, and it did not reduce the requirement that countries patrol wide areas of ocean. Rather, it increased the likelihood that nations undertaking such efforts could be sure that a violation had been committed.

The 1969 amendments' major innovative strategy, however, was to limit total discharges per ballast voyage to 1/15,000th of a tanker's total cargo-carrying capacity. The amendments did not change the then-current international law barring port states from the intrusive inspections of vessels that were needed to verify compliance, reserving that right for flag states. The Oil Companies' International Marine Forum (OCIMF), which had strongly pressed for adoption of these amendments, offered to provide a system by which private tanker inspectors would inspect whether a ship had clearly violated the total discharge limit.[14] Resolution A.391(X) forecast that the 1969 amendments would make detection of violations significantly easier. Compared to any of the discharge process criteria, this standard greatly reduced the detection demands. This new standard for the first time allowed states to detect violations and collect evidence in port after oil had been discharged rather than requiring them to monitor tankers' activities over wide areas at sea. The total discharge standard eliminated the need for states to rely on self-incriminating evidence from captains and allowed them to draw incontrovertible links between violations and perpetrators.

13. Article III(c)(i), *International Convention for the Prevention of Pollution of the Sea by Oil, 1954, as Amended in 1969,* 21 October 1969, reprinted in 1 I.P.E. 366, hereinafter cited as *OILPOL 54/69.*

14. See submissions to IMCO by OCIMF and ICS in MEPC II/4 (2 October 1974) and MEPC II/4/1 (14 November 1974) on the proposed "Voluntary In-Port Inspection."

MARPOL incorporated provisions identical or similar to those of the 1969 amendments, but also sought to improve the detection and identification of violations. As had been initially proposed in 1962 and recommended by an IMCO Assembly resolution in 1968, states were finally given the right to inspect ships in their ports for evidence of discharge violations.[15] Since port states had always been allowed to inspect ships' oil record books for evidence of discharge process violations, the new legal right was most important because it allowed states to verify compliance with the total discharge limit. More important, states assumed a general obligation to "cooperate in the detection of violations and the enforcement of the provisions of the present Convention, using all appropriate and practicable measures of detection and environmental monitoring."[16] In addition, parties assumed a specific obligation that they "should, to the extent they are reasonably able to do so, promptly investigate" all incidents in which visible traces of oil were observed.[17] Apparently these new enforcement obligations were expected to result in greater levels of enforcement by giving states a sense of legal obligation and by removing barriers to inspections that port states presumably felt were constraining their ability to enforce treaty provisions. Increased attention to enforcement reporting after 1973 was also expected to put substantial pressure on governments to increase enforcement, "since governments do not want to have a reputation for illegal or irresponsible behavior."[18]

All of these changes have been made to increase the likelihood that violations would be detected and their perpetrators identified. Although no single source provides reliable and consistent data on the detection and identification of discharge violations, even evidence from several sources fails to demonstrate that any of these changes, either collectively or individually, has had such an effect. Little change in detection effort or effectiveness was noticed after 1978, when the 1969 amendments took effect, or after 1983, when MARPOL took effect.

15. Article 6(5), *MARPOL 73/78*.
16. Article 6(1), *MARPOL 73/78*.
17. Annex I, Regulation 9(3) and 10(6), *MARPOL 73/78*.
18. M'Gonigle and Zacher, *Pollution, Politics, and International Law,* 336.

With regard to detection efforts and states' obligation under MARPOL to use "all appropriate and practicable measures" to detect violations, many developed nations such as the European states, Japan, Canada, and the United States have extensive in-port inspection programs. For example, the British began an extensive in-port inspection program in 1975, eight years before MARPOL entered into force.[19] In almost all cases, however, these in-port inspection programs have looked for certification and equipment violations, not for evidence of discharge violations. Until 1992 the Memorandum of Understanding on Port State Control (MOU) specifically excluded in-port inspections related to discharge violations from its mandate.[20] Only Germany has reported using in-port inspections to detect violations of discharge standards, using "no proof of the whereabouts" of sludge or other oil wastes as evidence of a discharge violation.[21]

Many developed states have undertaken aerial surveillance programs, although the size and quality of these programs is usually undocumented. The Netherlands began extensive aerial surveillance programs in 1975, but has decreased total flight time slightly since it peaked between 1980 and 1982. The United States developed and deployed an Airborne Oil Surveillance System plane in the mid-1970s for oil spill detection, but subsequently discontinued dedicated oil pollution surveillance flights.[22] The French began an aerial surveillance program in the late 1970s. In 1989 the Dutch and Belgians began a small cooperative aerial surveillance program. Australia, Denmark, Greece, the Federal Republic of Germany, Japan, Norway, and South Africa are also known to have used

19. James Cowley, "The International Maritime Organisation and National Administrations," *Transactions of the Institute of Marine Engineers* 101 (1989).

20. Secretariat of the Memorandum of Understanding on Port State Control, *Annual Report 1992* (The Hague: The Netherlands Government Printing Office, 1992), 24.

21. See, for example, German reports in MEPC 30/17 (20 July 1990) and MEPC 32/14 (13 November 1991).

22. The oil pollution system surveillance flights were made one of numerous other surveillance missions during the mid-1980s because of decreased funding and priority reassessment (Interview, Daniel Sheehan, U.S. Coast Guard, Washington, DC, 9 April 1992).

aerial surveillance to detect violations.[23] The most exhaustive analysis of enforcement data reported to IMO concluded that "it is likely that many Contracting Parties are not using all appropriate and practicable measures of detection . . . and, therefore, are not complying with Article 6(1) of the Convention."[24]

In contrast, of the few developing states that have reported to IMO, none has reported regular use of naval and aerial surveillance or in-port inspections. Especially in the case of expensive programs to detect violations at sea, whether by ship or by aircraft, developing states lack the necessary resources and "do not place high priority on this function in allocating scarce resources" that they do have.[25] As discharges are released predominantly on ballast voyages, the only point at which real

23. For a discussion of aerial surveillance programs reported to IMO, see Gerard Peet, *Operational Discharges from Ships: An Evaluation of the Application of the Discharge Provisions of the MARPOL Convention by Its Contracting Parties* (Amsterdam: AIDEnvironment, 1992), 11–12. For Japan, see MEPC XV/11/1 (6 January 1981). For the United Kingdom, see United Kingdom Department of the Environment, *Oil Pollution of the Sea* (Pollution Paper no. 20) (London, England: Her Majesty's Stationery Office, 1983), 23. For France, see MEPC X/13/6 (10 November 1978). On the U.S. Airborne Oil Surveillance System see MEPC IX/11/1 (2 March 1978) and MEPC 21/Inf. 9 (25 March 1985). A general discussion of enforcement that includes mention of Dutch, French, and German aerial surveillance programs among European Community states is contained in James McLoughlin and M. J. Forster, *The Law and Practice Relating to Pollution Control in the Member States of the European Communities: A Comparative Survey* (London: Graham & Trotman, 1982), chapter 6. On Dutch-Belgian cooperation in aerial surveillance see Belgian Marine Environmental Control, "Programme de Surveillance Aerienne de la Pollution Marine," unpublished paper, Brussels, Belgium, 1990; and Greenpeace, *Implementation of the Second North Sea Conference: Belgium* (Greenpeace Paper 11) (Amsterdam: Greenpeace International, 1989), 88. On a Danish program begun in 1989, see MEPC 29/Inf. 13 (8 February 1990). In 1990 the Helsinki Commission conducted an aerial surveillance exercise involving aircraft from Denmark, the Federal Republic of Germany, Finland, the German Democratic Republic, Poland, the Soviet Union, and Sweden; see "Seven Nations' Aircraft to Fly Patrol to Search for Ships Polluting Baltic Sea," *International Environment Reporter* 13 (August 1990).

24. Peet, *Operational Discharges,* 11–12 and Annex 10.

25. William T. Burke, Richard Legatski, and William W. Woodhead, *National and International Law Enforcement in the Ocean* (Seattle: University of Washington Press, 1975), 126. The authors conclude that "it is probably a mistake to believe that developing port states will play any consequential role in enforcement."

monitoring of compliance with the total discharge standard can be done is in oil-loading states. Only four of thirteen members of the Organization of Petroleum-Exporting Countries (OPEC) have signed MARPOL, and none has ever reported on its enforcement activity. Of the eight OPEC members that signed OILPOL, only Kuwait has provided an enforcement report of any type to IMO. While some developing states may be conducting in-port inspections but not reporting violations detected through this means, this seems unlikely.[26] Those states with the practical ability to monitor the 1/15,000th rule were not treaty parties, would be spending their own enforcement resources to keep other countries' coastlines clean, and would be making their own ports unattractive compared to those of states with less burdensome inspection procedures.

Until port states gained the right to conduct in-port inspections under MARPOL in 1983, detection of total discharge violations relied upon inspections by flag states or voluntary in-port inspections by oil company personnel at oil-loading terminals. No evidence exists that flag states even considered stationing inspectors in oil-loading port states to monitor compliance with the total discharge standard. OCIMF and the International Chamber of Shipping (ICS) sought to promote voluntary inspections and had proposed to make the resulting information available to oil-exporting states. During the 1970s, oil companies did conduct inspections of the quantities of oil wastes tankers were retaining on board. The program established for this purpose, however, was not the one originally proposed but never authorized by IMCO, but a nonpublic effort to evaluate how widely and effectively tankers were using new load on top (LOT) procedures that did reduce discharges, but were of most interest to oil companies because they conserved increasingly valuable oil. Oil-loading states showed "no interest" in taking the oil companies up on their offer of the results of these surveys.[27] In fact, the oil companies discontinued the survey program in 1978, the very year in which the total discharge limits took effect. Neither the oil companies and flag states under the 1969 amendments nor the oil-loading port states under

26. "The most serious" enforcement problem "has been the lack of interest on the part of the oil-exporting states to inspect tankers in their ports"; see M'Gonigle and Zacher, *Pollution, Politics, and International Law,* 338.

27. M'Gonigle and Zacher, *Pollution, Politics, and International Law,* 333.

MARPOL had incentives to actually expend the resources necessary to detect violations of the total discharge standard. Although detection of a violation of the total discharge standard necessarily involved the collection of information regarding the responsible tanker, since excessive discharges became illegal in 1978 no one has conducted the inspections that would make such information valuable as the first step in the deterrence process.

These facts suggest that, although states are enforcing the treaty, the means by which and degree to which they do so have not been influenced by treaty obligations to enforce. Most states simply have not rigorously enforced oil pollution regulations. The 1969 clean ballast and total discharge changes did not increase detection efforts. Neither did MARPOL's new obligations that states use all practical and appropriate means of enforcement significantly increase the detection of discharge violations. Besides, what evidence exists supports the notion that states that have developed aerial surveillance and in-port inspection programs might well have done so even without the MARPOL requirement: many programs began before the MARPOL requirement took effect, others began more than five years after the requirement took effect, and some have been discontinued despite the requirement. The oil companies discontinued their inspections just as the requirement gained legal force. Apparently the treaty's requirement has had little direct influence on decision making to undertake such programs.

While detection efforts have not been influenced by regulatory changes, these changes were also intended to improve the likelihood that such efforts would identify violations. National reports show that the number of alleged discharge violations that governments detected and reported to IMO for 1978 through 1980 was almost triple the number reported for 1975 through 1977. It is difficult to attribute these findings to increased detection, however, since there was a similar increase in the number of countries reporting and the average number of violations reported per country remained relatively constant throughout the period. Since 1983, while the number of reporting countries has remained comparable to the number reporting for 1978 through 1980, the total number of alleged violations reported and the average number per country have dropped. Dutch statistics provide more consistent figures on spills detected in the North Sea from 1969 to 1988. These figures count oil slicks detected, regardless of source, rather than discharge violations. Although they show a marked increase after 1977, the increase seems more likely to be due to the improved effectiveness of the aerial surveillance techniques begun in 1975 than to any effect of the clean ballast rule.

The Dutch have clearly identified ships as the cause of such spills in only 14 percent of cases each year. Authorities could identify the specific ship responsible in fewer than one-third of these cases, even though they knew a specific ship was involved. Improving detection technologies may ironically have made identification of violators more difficult: when the Dutch adopted radar that could detect spills at night in 1983, they detected more spills but identified the responsible ship in a smaller fraction of cases, because the radar could detect a spill but not a ship's name at a long distance.[28] The British fared only slightly better from 1978 to 1980 when they linked an average of only twenty-two percent of spills to ships. Although IMO has not adopted any of the regular proposals to "tag" oil shipments with cargo-unique compounds or to conduct detailed chemical analyses of spill samples to identify violators, these proposals attest to the ongoing obstacles that prevent authorities that detect discharges from identifying the responsible tankers.[29]

Although the quality of available data is low, the data that do exist suggest that the 1969 amendments to OILPOL and MARPOL resulted in little increase in states' success in detecting discharge violations and identifying the violators. Why did the various regulations fail? The clean ballast regulation failed to address the real source of the problem. States had dedicated few resources to detecting discharge violations because they had lacked incentives to do so, not because they had faced practical obstacles to doing so. The clean ballast regulation solved problems related to capacity, not to incentives. Detecting violations of the clean

---

28. N. Smit-Kroes, *Harmonisatie Noordzeebeleid: brief van de Minister van Verkeer en Waterstaat* (Tweede Kamer der Staten-Generaal: 17–408) (The Hague: Government Printing Office of the Netherlands, 1988).

29. On Swedish tagging, see MEPC I/Inf. 5/Add. 1 (4 March 1974); MEPC III/Inf. 15 (20 June 1975); and MEPC IX/17 (11 May 1978), 29. On the use of fingerprinting "with a view to facilitating the enforcement of the Convention," see MEPC I/6 (10 January 1974); and J. A. Butt, D. F. Duckworth, and S. G. Perry, eds., *Characterization of Spilled Oil Samples: Purpose, Sampling, Analysis and Interpretation* (London: John Wiley and Sons, 1985).

ballast regulation still required governments to undertake large and expensive surveillance programs that, under the best of circumstances, would leave most violations unobserved and most violators unidentified.

The total discharge limits had little impact on government enforcement because it imposed new rights and obligations on those that had little logical incentive, and had shown little historical incentive, to undertake the actions expected of them. Flag state governments never undertook in-port inspections. Oil company staffs initially undertook such inspections, but port states did not avail themselves of the results. After the total discharge regulation took legal effect, oil companies discontinued their surveillance programs, and oil-exporting states have had no incentives to undertake such programs since.

MARPOL's rules making detection efforts an obligation have also failed, because what new rights they provided were given to states with few incentives to enforce and the new obligations they did impose did not come with any new incentives for states to undertake the required efforts.

Improving Prosecution and Conviction

Even if these rule changes did not make detection and identification more likely, they did hold the potential to improve the prosecution and conviction of those violators that were identified. Modifications to the discharge standards have never dismantled the exclusive right of flag states to enforce regulations in their jurisdictions, which was criticized as a major obstacle to the effective prosecution and conviction of violators under the original rules of the 1954 convention. However, several countries have made proposals to improve the effectiveness of this system.

During the 1960s various proposals were made that were aimed at increasing the likelihood that violators would be prosecuted and convicted. A 1969 proposal to require flag states to prosecute violations referred to them by other states was opposed as an international legal infringement on a state's legal system.[30] The clean ballast and total discharge requirements of the 1969 amendments to OILPOL both promised to increase prosecution and conviction because they reduced the evidentiary problems that had plagued the rules of the 1954 convention.

30. M'Gonigle and Zacher, *Pollution, Politics, and International Law,* 224.

With respect to the clean ballast rule, "any sighting of a discharge from a tanker within 50 miles from land would be much more likely to be evidence of a contravention of the Convention."[31] It was hoped that photographic evidence linking a spill to an identified tanker would overcome the previous reluctance to prosecute, and failure to convict, alleged violators.[32] The total discharge rules could have also made prosecution and conviction more likely since, upon detecting "gross contraventions" of the discharge limit, oil company staff were to send evidence to the port state for further referral to the flag state for prosecution.[33]

Numerous enforcement improvements were also proposed at the 1973 conference, mainly by the United States, Canada, Australia, and New Zealand, including a radical proposal to allow a state to prosecute the owners of any violating ship in its ports regardless of where it had committed a violation.[34] This proposal was defeated by a vote of twenty-five to sixteen, but MARPOL eventually did include two basic improvements to its enforcement provisions. Both strengthened the obligations to refer and to prosecute discharge violations. MARPOL obligated states not only to detect violations, but to use "adequate procedures for reporting and accumulation of evidence."[35] States now were required, rather than merely allowed, to refer violations detected outside territorial seas to the flag state.

Subsequent IMCO Assembly resolutions reiterated the need for port and coastal states to provide "a ready flow of information concerning alleged contraventions in order to facilitate enforcement by the [flag state]."[36] One resolution recommended standard procedures for investi-

31. Resolution A.391(X) (14 November 1977), Annex, par. 5.

32. Resolution A.391(X) (14 November 1977), Annex, par. 6; and M'Gonigle and Zacher, *Pollution, Politics, and International Law,* 328.

33. As an example, Shell Oil Company would report to Saudi Arabian port authorities that a Panama-registered ship that Shell owned or was chartering had violated the discharge standards. The Saudi Arabian authorities were then to take steps to develop further evidence and turn this over to Panamanian authorities for prosecution.

34. M'Gonigle and Zacher, *Pollution, Politics, and International Law,* 231.

35. Article 6(1), *MARPOL 73/78.*

36. Resolution A.391(X) (14 November 1977). See also Japanese proposals for minimum acceptable evidence in MEPC XI/14 (19 March 1979).

gating and referring observed violations to flag states under the OILPOL agreement. A similar 1983 resolution noted the claims of flag states that referrals often lacked information adequate to allow them to proceed with prosecution. The resolution recommended more stringent measures for detection and provided an itemized list of types of evidence that referring states should provide to flag states to facilitate prosecution.[37]

Therefore, a series of changes have been made in an effort to improve the likelihood and success of the prosecution of discharge violations. Any impact of these changes should be evident in an increase in the number of states referring violations to flag states, an increase in the share of detected violations actually prosecuted, or a decrease in the share of cases turned down for insufficient evidence.[38] However, IMO statistics show that enforcement by port, coastal, and flag states has remained largely unaffected by these efforts at improvement. Although there are insufficient data from earlier periods with which to compare more recent data, the latter show port and coastal states have generally failed to prosecute most cases. British data for 1978 through 1980 show that the owners of only nine of fifty-five tankers caught deliberately discharging (16 percent) were successfully prosecuted.[39] IMO reports show that port and coastal states did not prosecute an average of 36 percent of cases involving violations in territorial seas between 1983 and 1990 because they lacked sufficient evidence. During this same period, port and coastal states successfully fined the shipowners in less than 28 percent of discharge cases.[40] In most cases, however, international law required states identifying discharges to refer them to the flag state.

37. Resolution A.542(13) (17 November 1983).

38. Both IMCO and the tanker industry maintained that governments could not begin enforcing these rules until after entry into force. See Resolution A.391(X) (14 November 1977); and William Gray, "Testimony," in U.S. House of Representatives, Committee on Government Operations, *Oil Tanker Pollution—Hearings: July 18 and 19, 1978* (95th Congress, 2nd session) (Washington, DC: GPO, 1978).

39. Peet, *Operational Discharges*, tables 8 and 9, 13.

40. United Kingdom Royal Commission on Environmental Pollution, *Eighth Report: Oil Pollution of the Sea* (London: Her Majesty's Stationery Office, 1981), 195.

Table 5.1
Successful flag state prosecutions of violations of oil discharge regulations as a percent of referrals

Referring country	Period	Cases referred	Convicted by flag state	Percent
Canada	1967–1977	80	17	21
France	1968–1975	165	17	10
France	1976–1983	202	48	24
Japan	1978–1979	24	1	4
Japan	1980	34	5	15
Netherlands	1978–1980	37	1	3
United Kingdom	1976–1981	71	9	13
Average	1967–1983	613	98	16
All countries				
	1983	30	14	46
	1984	91	24	26
	1985	105	7	7
	1986	132	7	5
	1987	61	4	7
	1988	105	12	11
	1989	125	15	12
	1990	84	12	14
Average	1983–1990	733	95	13

Sources: For 1983–1990, Gerard Peet, *Operational Discharges from Ships: An Evaluation of the Application of the Discharge Provisions of the MARPOL Convention by Its Contracting Parties* (Amsterdam: AIDEnvironment, 15 January 1992). For 1967–1983, adapted from various MEPC documents cited in Paul Stephen Dempsey, "Compliance and Enforcement in International Law: Oil Pollution of the Marine Environment by Ocean Vessels," *Northwestern Journal of International Law and Business* 6 (1984); data for Canada 1967–1977 from MSC/MEPC/WP.12 (17 October 1977); and data for France 1967–1983 from MEPC 21/16/3 (30 January 1985).

The more specific clean ballast criteria should have led states to refer more cases since they could more easily produce evidence sufficient to induce flag states to prosecute. In incidents presumably beyond territorial seas, port and coastal states reported leaving action to the flag state in 1,077 of 1,335 cases (80 percent). Although the reporting of referrals was inconsistent and the number of referring countries increased, the average number of referrals per reporting state was, if anything, lower in the 1980s than it was in the 1970s (see table 7.1, infra). These referrals also appear not to have improved the problem of getting flag states to follow up on prosecutions. Complaints to IMO of flag states' inaction document low conviction rates by flag states. Of cases referred to flag states by a variety of port states, the flag states themselves prosecuted only 16 percent of all pre-1983 referrals for which information was available, turning down many of them for lack of sufficient evidence (see table 5.1).[41] Since MARPOL took effect in 1983, an average of only 13 percent of referrals to flag states have been successfully convicted and fined. Some 15 to 20 percent of the cases have not even been prosecuted because of lack of evidence.[42] A 1989 study of three hundred referrals by North Sea states found that flag states had taken action on only 17 percent.[43] These figures suggest that insufficient evidence and disincentives to prosecution make convictions no more likely now than previously. As a 1985 French analysis put it, "only a few States in fact follow up the reports that are transmitted to them."[44] Even after the 1969 amendments to OILPOL and MARPOL took effect, insufficient evidence has remained a major problem and successful prosecutions have remained elusive.

Although the clean ballast rule was defined so that photographic evidence should have been sufficient evidence of a violation, states have

41. Paul Stephen Dempsey, "Compliance and Enforcement in International Law: Oil Pollution of the Marine Environment by Ocean Vessels," *Northwestern Journal of International Law and Business* 6 (1984).

42. Peet, *Operational Discharges,* tables 11 and 12, 17–18.

43. Marie-Jose Stoop, *Olieverontreiniging door Schepen op der Noordzee over de periode 1982–1987: Opsporing en Vervolging* (Amsterdam, The Netherlands: Werkgroep Noordzee, July 1989).

44. MEPC 21/16/3 (30 January 1985).

never agreed to accept photographic evidence of an oil slick as legal evidence of a violation.[45] A major obstacle is that detected violations cannot be linked reliably to the tankers responsible. Evidence from aerial surveillance programs is not always sufficient to convince port state prosecutors to pursue a case, let alone the prosecutors in a flag state many miles away from the scene of the discharge.[46] At a 1991 court simulation conducted by European maritime authorities, a Belgian official playing the role of defense attorney successfully argued that the heaviness of North Sea tanker traffic precluded the finding that photographs of a tanker in the immediate vicinity of an illegal discharge were evidence that the tanker was responsible.[47] Also, OILPOL and MARPOL reinforce each state's right to use its own legal standards to determine whether "sufficient evidence is available to enable proceedings to be brought in respect of the alleged violations."[48] Flag states, therefore, face no binding obligation to prosecute violations referred by other states. A flag state's incentives generally run counter to vigorous enforcement of violations, and the claim that there is insufficient evidence of a violation provides one means of avoiding action. As already noted, the total discharge standard could have improved this state of affairs, but flag and oil-loading port states never collected the necessary information and oil companies discontinued collecting such information after the 1969 amendments took effect.[49] No evidence exists that the total discharge provisions have ever been used as a basis for referral or prosecution.[50]

These efforts did not increase detections and referrals because they failed to address the real obstacles to prosecution and conviction. The

45. Ton IJlstra, *Enforcement of MARPOL: Deficient or Impossible, Marine Pollution Bulletin* 20 (December 1989).

46. See Cowley, "The International Maritime Organisation," 138–139, on the difficulties of prosecution even with aerial surveillance evidence.

47. Interview, Ronald Carly, Ministry of Transportation—Marine Division, Brussels, Belgium, 10 June 1991.

48. Article IV(1), *OILPOL 54.*

49. Oil Company representatives were quick to point this out when, in response to congressional demands, they released the results of these surveys; see William O. Gray, "Testimony." On the failure of the industry to provide the information, see M'Gonigle and Zacher, *Pollution, Politics, and International Law,* 228.

50. Interview, Nigel Scully, British delegate to IMO, London, 5 July 1991.

inherent difficulty of identifying perpetrators often made conviction unlikely, a problem that standards governing the discharge process never overcame; identifying violations of the total discharge rules could have overcome this problem, but authorities never inspected for them. Port states often had little basis on which to prosecute or refer cases, even though referral was now mandatory. Incentives for such referrals were further reduced by the knowledge that prosecution depended on the flag state's judgment that the evidence was sufficient. Although IMO had produced standards for the evidence provided in referrals to make prosecution more likely by flag states, it never made meeting those requirements sufficient to obligate such prosecution, and doing so would have been unlikely to increase prosecution in any event.

The drafters of MARPOL sought to increase the referral and prosecution of violations, assuming that international legal obligations could overcome states' material disincentives to expend resources to investigate cases in which prosecution was subject to unilateral veto by the flag state. Flag states had already shown that they were unlikely to follow up on cases in any event. MARPOL failed to provide any new rights to states or to address the fundamental source of the problem because of political resistance to infringements on the sovereign legal and judicial systems of member states. It has been seen that "nations both view their own [legal] systems jealously and are suspicious of the systems employed by others."[51] In 1962 British enforcement proposals were rejected as "incompatible" with European legal systems.[52] As late as 1990, the MOU Secretariat noted that successful prosecution remained a "very complicated matter" because different "legal standards for collection and international exchange of evidence to courts" remain an unresolved problem.[53] The political inability of states to agree to give greater rights of prosecution to the port states with incentives to prosecute has meant that efforts to increase referral and prosecution have remained unsuccessful.

51. Burke, Legatski, and Woodhead, *National and International Law Enforcement*, 132.

52. M'Gonigle and Zacher, *Pollution, Politics, and International Law*, 222.

53. Secretariat, *Annual Report 1990*, 20.

Increasing Penalty Levels

Besides attempting to increase the probabilities of detecting and convicting violators, efforts have been made to improve the deterrence of the discharge provision's compliance system by inducing governments to impose stiffer penalties in those cases in which convictions have been handed down. Although nations at the 1962 conference rejected a proposal that would have set a target level for penalties "as infringing on the autonomy of states' legal systems," they adopted a U.S. proposal requiring states to make penalties "adequate in severity" to prevent discharge violations.[54] Although the imposition of serious penalties was not required in the original 1954 convention, nations had recognized as early as the 1935 draft convention that the "inherent difficulty" of detecting violations of the convention's discharge provisions required that the parties "seriously consider the question of imposing penalties which will be sufficient to act as a deterrent in connection with such offenses."[55]

Since 1962, IMO treaties and resolutions have consistently tried to induce states to impose penalties large enough to deter tanker captains from committing violations in the face of the low probability of being detected, identified, prosecuted, convicted, and fined. Frequent IMCO Assembly resolutions as well as MARPOL itself have all reiterated the need for states to increase the penalties imposed on convicted violators in order to deter others.[56] Since no agreed-upon definition of *adequate* existed, states could still set fines as they pleased without violating the treaty. The rule provided no new right and established no clear obligation. Rather, the 1962 language and subsequent resolutions were intended to alter norms regarding appropriate fine levels in the face of levels

54. Article VI(2), *OILPOL 54/62;* M'Gonigle and Zacher, *Pollution, Politics, and International Law,* 222; and United States Senate, "Message from the President," 40.

55. Recommendation 3, League of Nations Committee of Experts on Pollution of the Sea by Oil, *Draft Final Act of the International Conference Relating to the Pollution of the Sea by Oil* (C.449.M.235.1935.VIII) (Geneva: League of Nations, 1935).

56. See Resolution A.151(ES.IV) (26 November 1968); Resolution A.153(ES.IV) (27 November 1968); Resolution A.412(XI) (15 November 1979); Resolution A.499(XII) (19 November 1981); and Article 4(4), *MARPOL 73/78.*

that were clearly inadequate to solve the problem. To use embarrassment to put such normative pressures on states reporting low fine levels, nations agreed in 1962 to require states to report on penalties actually imposed.[57] The United States, for one, considered that including such reports in the secretary-general's annual reports "strengthened considerably" the penalty provisions.[58]

This continuing effort to increase fines has met with little success. As in other areas, the lack of any data before countries began reporting on treaty-related activities in 1975 requires us to rely on anecdotal evidence. Given the absence of reports, the mechanism of shaming countries to increase their fine levels was absent. The IMCO Assembly's long discussions in 1968 on how to make OILPOL more effective included extensive attention to fine levels. Therefore, it is unlikely that things had changed much since the 1961 survey, whose responses were evidence of low fine levels.[59] Strictly speaking, OILPOL only required states to implement statutes with fines that were adequate in severity, although the goal was clearly to raise fine levels. Although statutory fine levels are not available for most countries, a few were increased somewhat during the 1970s: British and Canadian maximum fines rose from £1,000 to £50,000 and $5,000 to $100,000, respectively, in the early 1970s, but U.S. maximums were only $10,000 as of 1979.[60] Even these levels were rarely imposed, however. In terms of fines actually imposed, by the time that the first data were reported in 1975, fines were still quite low by any standard. Table 5.2 shows that average nominal fines have exceeded the 1975 level of $4,074 in only two years; adjusted for inflation, fines have never exceeded this amount.

The effort to increase fines failed for four reasons. First, the efforts to increase penalties did not remove any of the legal or practical barriers to the imposition of high penalties. A state had neither more nor less legal authority or practical ability to impose high fines after the adoption of the new 1962 language or the various resolutions than it did before.

57. Article VI(3), *OILPOL 54/62*.

58. United States Senate, "Message from the President," 49.

59. See Resolution A.151(ES.IV) (26 November 1968); Resolution A.153(ES.IV) (27 November 1968); and Resolution 155(ES.IV) (27 November 1968).

60. M'Gonigle and Zacher, *Pollution, Politics, and International Law,* 228.

Table 5.2
Average fines imposed by states reporting to IMO

	1975	1976	1977	1978	1979	1980
Number of fines	130	119	162	177	370	142
Value of fines (000s of then-year U.S.$)	530	226	455	352	870	532
Average value of fines (then-year U.S.$)	4,074	1,902	2,807	1,987	2,351	3,743
Average value of fines (constant 1982 U.S.$)	6,613	3,000	4,286	2,707	2,737	3,928

Sources: Paul Stephen Dempsey, "Compliance and Enforcement in International Law: Oil Pollution of the Marine Environment by Ocean Vessels," *Northwestern Journal of International Law and Business* 6 (1984), 488; Gerard Peet, *Operational Discharges from Ships: An Evaluation of the Application of the Discharge Provisions of the MARPOL Convention by Its Contracting Parties* (Amsterdam: AIDEnvironment, 15 January 1992), annex 15 and table 13, p. 21; and Producer Price Index for Crude Materials from U.S. Bureau of the Census, *Statistical Abstract of the United States: 1991* (Washington, D.C.: GPO, 1991), 481.

Besides, the language incorporated has never been specific. *Adequate fines* has never been defined, and a specific target fine level has never been established. A state could not know if it was complying with the new provision even if it had wanted to comply. The obligation was both weak, requiring the imposition of high fines on paper but not necessarily in practice, and also nonspecific.

Another reason these efforts failed was that they left each government's incentives unaltered. The IMCO Secretary-General could not shame countries into raising fine levels by publishing national reports on them, because it failed to receive any such reports until 1975. Until the standardized reporting formats were developed in 1985, many reports IMCO did receive failed to report fine levels.[61] Even now, countries often report fines in local currency, and IMO has not converted them to analyze time trends across countries. Therefore, evaluating a state's compliance with

61. Dempsey, "Compliance and Enforcement," 488.

Table 5.2
(continued)

1981	1982	1983	1984	1985	1986	1987	1988	1989	1990
19	na	29	41	297	593	4	439	478	544
103	na	25	97	219	360	3	292	448	3,350
5,393	na	876	2,358	737	607	840	664	937	6,158
5,236	na	865	2,278	769	692	896	692	909	5,865

the "adequate in severity" provision regarding penalties has been essentially impossible.

The requirement of fines adequate to deter violators also ignored the domestic constraints on fine levels. In many countries independent judiciaries, not executives, set the standards for legal penalties. Given the low probability that a violator would be detected, prosecuted, and convicted, fines large enough to deter others would have to be disproportionate to the actual harm of the illegal discharge.[62] Although deterrence often enters the calculus of setting fines, most legal systems resist imposing disproportionate penalties against a few convicted violators to compensate for the enforcement system's inability to make detection and successful prosecution likely. The new rules required states to impose penalties at levels that ran counter to the major domestic determinants of penalties. The decision to require states to impose high fines for discharge violations was the first enforcement improvement that was adopted precisely because it skirted controversial jurisdictional issues raised by other efforts to remedy detection and prosecution problems.

Summary of Improvements in Enforcement

A wide array of policies have attacked each of the three phases of enforcement of the OILPOL and MARPOL discharge standards. Yet it

62. For an economic analysis of penalty levels, see Gary S. Becker, "Crime and Punishment: An Economic Approach," *Journal of Political Economy* 76 (March/April 1968).

is hard to find much evidence of substantial improvement in the detection of violations or in the prosecution or penalization of violators. Whereas significant enforcement efforts are being made, as in the area of aerial surveillance in some countries, these efforts have little correlation with the various policy changes, suggesting that other factors are at work. In many cases, like those of fine levels and flag state conviction rates, the expected changes in behavior simply are not evident.

The clean ballast provision failed to improve enforcement primarily because the requirements for effective enforcement far exceeded the capacity of even an industrialized country. Even the environmentally committed United States conducted little surveillance beyond its ports during the 1970s.[63] Like the original 100 ppm standard that it replaced, the clean ballast rule placed large demands on government enforcement resources. Enforcement still required costly aerial or naval surveillance of large areas of ocean, while tankers could still easily escape detection by discharging at night or in bad weather. At the same time, the benefits of conducting surveillance beyond the territorial sea remained few because detected violations would have to be referred to a flag state, which often would not prosecute. The nature of the crime also left many of the practical obstacles to detecting the violation and identifying and successfully prosecuting the violator unaddressed. The clean ballast requirement provided no new incentives for states to undertake enforcement, making it more likely only that detection efforts would result in actual detection of violations.

Compared to the clean ballast standard, the total discharge standard greatly reduced the monitoring demands placed on governments and undoubtedly would have made prosecutions more successful by ensuring identification of the perpetrators of violations. However, it failed to increase enforcement efforts because, of the actors with the practical ability and legal authority to monitor tankers' compliance with the standard—flag states, oil companies, and oil-exporting governments—none had incentives to do so.

MARPOL's adoption of explicit requirements for states to detect and prosecute discharge violations, as well as earlier efforts to increase fine levels, failed because none of them successfully established a mechanism

---

63. M'Gonigle and Zacher, *Pollution, Politics, and International Law*, 332.

for increasing the incentives of governments to undertake these activities. No process existed for converting the requirements on paper into practice. Undertaking detection, prosecution, or penalization was neither easier nor more rewarding after the treaty change than before. Failing to undertake these activities had no consequences in shaming or embarrassing the violators.

Most important, these strategies did not succeed because even collectively they addressed only part of the problem. The basic deterrence chain has many links, from detection and identification through prosecution and conviction to penalization. Failure to induce a government to undertake one of these processes increases the demands on other parts of the compliance system. In short, the enforcement of discharge standards involved a high-maintenance deterrence system that presented a variety of pathways to failure and only a very few to success.

Enforcement of Equipment Standards

The general failure of the numerous efforts to enhance the enforcement of discharge standards provided one of the major stimuli for the 1973 conference that drafted MARPOL. The United States and some other countries viewed many of the problems of enforcing the discharge standards as inherent. Modifications to that enforcement system would not work; improvement demanded that changes be made in the underlying primary rules themselves. If low levels of enforcement of and compliance with the discharge standards were "a function of the rule," then a qualitatively different enforcement system was needed.[64] Just as the initial discharge standards provided a structure that largely determined the type and success of efforts to encourage enforcement, the adoption of equipment standards created a new enforcement structure that defined how and how much countries monitored and sanctioned violations.

Various proposals at the 1973 conference—by the United States and other environmentalist states and by developing coastal states seeking to increase their enforcement jurisdictions in nonenvironmental realms— were intended to broaden port states' detection and prosecution rights.

64. Roger Fisher, *Improving Compliance with International Law* (Charlottesville, VA: University Press of Virginia, 1981).

Port state enforcement was rejected by a vote of twenty-five to sixteen, with strong opposition from maritime states concerned about freedom of navigation, including European, Soviet bloc, and flag-of-convenience states. Conference delegates did, however, establish new construction and equipment standards, most notably requiring segregated ballast tanks (SBTs) on large new tankers, which would dramatically increase enforcement. The 1978 protocol conference added to the convention equipment requirements for new and existing tankers, including the installation of crude oil washing (COW) equipment, and refined the compliance system before it entered into force as MARPOL 73/78 in 1983.

In fact, the compliance system underlying the equipment standards relies far less on deterrence and enforcement than do the discharge standards. MARPOL established a system that created obstacles to noncompliance in the first place, using deterrence mainly as a backup mode of discouraging noncompliance. Some of the features of the system were inherent in the equipment standards themselves, while others grew out of specific supporting provisions agreed to at the 1973 and 1978 conferences.

By regulating a transaction involving many actors, the equipment regulations limited the monitoring burden governments faced. In fact, the primary burden of monitoring fell on nonstate actors. New construction or retrofitting required the cooperation and approval of a shipbuilder, a classification society, and an insurance company. Each of these agents was, at a minimum, likely to demand an explanation of a request to omit legally required equipment. Even if they would not prevent the construction or block the tanker's ability to trade, the foreknowledge that they might and the need for an explanation undoubtedly dissuaded many owners from even making such requests.

MARPOL reinforced these processes inherent in regulating the tanker procurement transaction by also requiring flag states, or classification societies nominated by them, to inspect tankers for required equipment during construction or retrofitting. Regular inspections were to be undertaken every five years thereafter, supplemented by intermediate and unscheduled inspections. After the more extensive surveys, a tanker received an International Oil Pollution Prevention (IOPP) certificate that was valid for up to five years.[65] This system piggybacked on the existing inspection system used by classification societies during construction and retrofits for adherence to safety and other regulations. MARPOL also required classification society and flag state surveyors to "immediately ensure that corrective action is taken" if a ship were found to be violating the equipment requirements. These inherent and constructed elements of the compliance system reduced the number of potential violators and allowed governments to shift the monitoring burden to classification societies. Just as important, they allowed for the monitoring of behaviors that would lead to violation before a violation was actually committed.

Governmental monitoring served mainly to reinforce this private but treaty-authorized system. To keep nongovernmental surveyors and the tanker owners honest, MARPOL gave each port state the right to inspect an IOPP certificate and the ship itself if it had "clear grounds for believing that the condition of the ship or its equipment [did] not correspond substantially with the particulars of that certificate." IMO provided guidelines for these inspections the month after MARPOL took effect.[66] Unlike discharge violations, equipment violations were easily detected in port even long after the violations had been committed. And a detected equipment violation was incontrovertible, requiring no further investigation by the port state. Equally important, if an equipment deficiency was found, the treaty required the port state to "take such steps as will ensure that the ship shall not sail until it can proceed to sea without presenting an unreasonable threat of harm to the marine environment."[67] Besides requiring port states to detain noncompliant ships, the treaty explicitly reminded these states of their existing right to deny such ships entry to their ports. By legalizing detention, the drafters of MARPOL hoped to make the prevention of equipment violations more credible. As an administrative sanction, detention did not threaten flag states' legal systems. Imposing opportunity costs on a tanker raised none of the sovereignty issues of legal prosecution and fines. From a negotiation standpoint, paradoxically the imposition of such costs made flag states less likely to

---

65. Annex 1, Regulation 4, *MARPOL 73/78*.

66. Resolution A.391(X) (14 November 1977); and Resolution A.542(13) (17 November 1983).

67. Article 5(2), *MARPOL 73/78*.

object to the practice. From a practical standpoint, these administrative sanctions eliminated the legal obstacles that hindered successful conviction and penalization of those responsible for discharge violations.

Before MARPOL took effect, the maritime authorities of fourteen European states negotiated the Memorandum of Understanding (MOU) on Port State Control to enhance in-port enforcement of various IMO conventions.[68] The MOU coordinated efforts to inspect IOPP certificates and the existence and condition of required equipment and required each country to inspect 25 percent of the ships entering its ports. The MOU's centralized computer system in St. Malo, France, provided a common data base linked to national inspectorates that allowed states to avoid reinspections and focus their inspection efforts on tankers most likely to be in noncompliance, that is, those for which recent information was unavailable or that had shown recent deficiencies. Although the secretariat does not release information on individual states' inspection rates, the MOU's executive body, the Port State Control Committee, meets twice a year to discuss enforcement and the secretariat produces annual reports summarizing enforcement statistics across countries.[69]

The new primary rules and the supporting institutions had that "essential virtue" of any compliance system, namely enforceability; but the experience with total discharge standards attests to the fact that enforceable rules are not always well enforced.[70] Data on the availability and accuracy of IOPP certificates provide insight into the response of private and governmental actors to the new system. Data from both IMO and the MOU document that the absence of or discrepancies in IOPP certificates have fallen off dramatically since they were first required after MARPOL's entry into force in 1983. In both data sets, the total number

68. Secretariat of the Memorandum of Understanding on Port State Control, *The Memorandum of Understanding on Port State Control* (information pamphlet) (The Hague: The Netherlands Government Printing Office, 1989). Member Maritime Authorities include Belgium, Denmark, Finland, France, Germany, Greece, Ireland, Italy, the Netherlands, Norway, Portugal, Spain, Sweden, and the United Kingdom. See chapter 3.

69. George Kasoulides, "Paris Memorandum of Understanding: A Regional Regime of Enforcement," *International Journal of Estuarine and Coastal Law* 5 (February 1990).

70. M. P. Holdsworth, "Convention on Oil Pollution Amended," *Marine Pollution Bulletin* 1 (November 1970), 168.

Table 5.3
IOPP discrepancies[a]

Reported to the IMO Secretariat

Reporting country	1984	1985	1986	1987	1988	1989	1990
Australia					9	4	7
Bulgaria				38	24	12	9
China			70	34	73	79	91
Egypt				48	52	29	
Germany[b]		87	34	24	51	33	14
Greece					14	10	1
Hong Kong	0		0	0	0	0	0
Israel		0	0				
Italy	47	64	19				
Japan		0	0	0	0	0	1
Netherlands			92				
Norway	6	12	13	10	14	18	20
Sweden			15				
United Kingdom	82	132	116	75	50	35	43
United States	181	64	175	35	29	24	22
Total discrepancies reported	316	359	534	264	316	244	208
Number of reports	5	7	11	9	11	11	10
Average number per report	63	51	49	29	29	22	21
Number of ships inspected	3,602	14,610	21,879	32,332	29,957	27,444	35,243
As % of ships inspected	8.8	2.5	2.4	0.8	1.0	0.9	0.6

Reported to the MOU Secretariat

All MOU member states	1984	1985	1986	1987	1988	1989	1990
Total discrepancies reported	828	652	572	407	332	265	317
Number of reports	14	14	14	14	14	14	14
Average number per report	59	47	41	29	24	19	23
Number of ships inspected	7,686	7,879	8,721	10,337	8,382	9,164	9,842
As % of ships inspected	10.8	8.3	6.6	3.9	4.0	2.9	3.2

Sources: Various MEPC documents from 1984 to 1992; and Secretariat of the Memorandum of Understanding on Port State Control, *Annual Report* (The Hague: The Netherlands Government Printing Office, 1984–1990).

[a]"Discrepancies" can include the IOPP certificate's being unavailable, on board equipment's not matching the certificate, or on board equipment's not functioning. Countries include only those providing reports using IMO's mandatory reporting format. Blank spaces indicate years in which the country did not report; zeros indicate reports indicating no discrepancies.

[b]Combines reports from the Federal Republic of Germany and the German Democratic Republic for the years prior to 1990.

of discrepancies, the average number per report, and the percentage of inspected ships with discrepancies has dropped dramatically and steadily from highs of 9 to 10 percent in 1984 to recent lows of 1 to 3 percent (see table 5.3). Since IOPP certificate availability and accuracy are simple to verify, these trends show that most flag states and classification societies were conducting the surveys and inspections necessary to provide tankers with accurate certificates as early as 1984 and have considerably improved their procedures since then. The low—and declining—number of IOPP discrepancies suggests that flag states and classification societies do survey essentially all ships, and increasingly do it accurately.

Available data also demonstrate that several MARPOL signatories, though by no means all, were inspecting for equipment violations. The fourteen European MOU states made pollution prevention part of their standard inspection procedures and have stressed it in their training programs.[71] Although the MOU data aggregate discrepancies identified by all countries, it seems likely that all countries regularly checked IOPP certificates during inspections. Of the fifteen states that have reported to IMO, including seven MOU states, only three have reported finding no IOPP discrepancies, presumably reflecting a failure to inspect for such discrepancies rather than an actual absence of discrepancies. The fact that five countries besides the United States and MOU countries have reported IOPP discrepancies to the IMO Secretariat attests to the fact that almost all reporting countries have undertaken equipment inspections.[72] Japan, Canada, Russia, and Poland all have port state inspection programs. Ten Latin American states signed an MOU-like agreement in November 1992.[73] Of course, many more countries have signed MARPOL than have reported inspections, and the quality and frequency

71. Secretariat, *Annual Report 1990,* 24.

72. As I develop more fully elsewhere, the failure to report to IMO of seven states that regularly inspect and report to the MOU demonstrates that non-reporting cannot be assumed to be equivalent to noncompliance with enforcement requirements; see Ronald B. Mitchell, "Eliciting Reporting under Environmental Treaties," paper presented at the Northeast International Studies Association Conference, Providence, RI, November 1992.

73. *Acuerdo de Viña del Mar: Acuerdo Latinoamericano sobre Control de Buques por el Estado Rector del Puerto,* 5 November 1992 (original signatories include Argentina, Brazil, Chile, Colombia, Ecuador, Mexico, Panama, Peru, Uruguay, and Venezuela); and Secretariat, *Annual Report 1992,* 10–14.

of IOPP inspections undoubtedly vary widely. Nevertheless, inspections for MARPOL-required equipment do appear to have become a part of the inspections of more than twenty MARPOL states.

The MOU itself has played an important role in facilitating such inspections. The content of German, Dutch, French, and British reports to IMO in the 1970s suggests that these states already were conducting port state enforcement before the MOU took effect. Most appear to have increased their efforts since then, however. Data since 1976 show that the number of British in-port inspections increased from fewer than 1,800 prior to 1979 to over 2,900 in every year since 1981.[74] The dramatic increases in 1980 and 1981 correspond well to British involvement with the 1980 Hague MOU and its successor, the 1982 Paris MOU. Although domestic politics may partially explain why the United Kingdom increased its inspection rate in this period, it seems unlikely that these pressures would have produced such a dramatic and sustained increase in the absence of enforcement by other states.

MOU figures for ships inspected show a steady increase, from 6,325 representing 16 percent of ships entering MOU ports in 1982/3 to 9,842 ships representing almost 25 percent in 1989 (see table 5.4). The total number of inspections has increased from 8,839 in 1982/3 to almost 14,000 in 1990. Although the United Kingdom is the only state to consistently meet the MOU's 25 percent inspection rate requirement, the Netherlands, Greece, and Finland have also had relatively high inspection rates.[75] In contrast, Denmark, France, the Federal Republic of Germany, Ireland, and Italy have had moderate inspection rates (15 to 20 percent of ships entering their ports), but have shown no dramatic increase. Belgium, Norway, Portugal, Spain, and Sweden started with low inspection rates in the early 1980s and have significantly increased their inspection rates since. Whether these increases are due to improved reporting or to increases in the numbers of inspections is unclear, unfortunately.[76]

74. Cowley, "The International Maritime Organisation."

75. See Secretariat of the Memorandum of Understanding on Port State Control, "MOU Inspection Figures," Document PSSC17/15 (1991).

76. Although MOU reporting has been better than IMO enforcement reporting, some evidence indicates that dramatic increases in the percentages of ships inspected reflect increased reporting of inspections as well as increased numbers of inspections. This appears to be true for Belgium, for example: one source

Table 5.4
In-port inspection results

	1982/3	1983/4	1984/5	1984
Reports to the IMO Secretariat				
Countries reporting				5
Number of ships inspected				3,602
Average number per report				720
Reports to the MOU Secretariat				
Countries reporting	14	14	14	14
Number of ships inspected	6,325	7,342	7,665	7,686
Average number per report	452	524	548	549
As % of ships entering MOU ports[a]	16	16	17	17
Total number of inspections[b]	8,839	9,847	10,044	10,227
Average number per report	631	703	717	730

Sources: Various MEPC documents from 1984 to 1992; and Secretariat of the Memorandum of Understanding on Port State Control, *Annual Report* (The Hague: The Netherlands Government Printing Office, 1983/84–1990).

[a]The MOU Secretariat introduced new procedures for estimating the number of ships entering MOU state ports in 1988. The figures here use a conversion ratio (of 1.19) between the old and new estimating techniques. See Secretariat for the Memorandum of Understanding on Port State Control, *Annual Report 1988,* (The Hague: The Netherlands Government Printing Office, 1988), 19.

[b]Number of inspections exceeds number of ships inspected because some ships are inspected more than once.

shows that from 1985 to 1988 Belgian authorities inspected the oil record books of 704, 863, 827, and 949 ships, respectively. During this same period the number of Belgian inspections reported to the MOU rose from 2.5 percent of ships in port in 1985 to 13.5 percent, 14.2 percent, and 14.2 percent in 1986, 1987, and 1988, respectively. Given the implausible view that inspections of ship features other than the oil record book increased dramatically between 1985 and 1986, these figures suggest that many inspections were not reported in 1985. See Greenpeace, *Implementation of the Second North Sea Conference: Belgium,* 82–83.

Table 5.4
(continued)

1985	1986	1987	1988	1989	1990
7	11	9	11	11	10
14,610	21,879	32,332	29,957	27,444	35,243
2,087	1,989	3,592	2,723	2,495	3,524
14	14	14	14	14	14
7,879	8,721	10,337	8,382	9,164	9,842
563	623	738	599	655	703
17	19	19	18	21	23
10,417	11,740	11,451	11,224	12,459	13,955
744	838	817	801	889	996

Although it is difficult to confidently assess the significance of these numbers, the MOU appears to have led member states to maintain medium to high inspection rates or to raise low ones, even though most fail to meet the 25 percent target.[77] The absence of similar increases in other enforcement activities and the absence of plausible independent sources for such changes make it unlikely that the increases in inspections would have occurred without the MOU.

Governments are not merely monitoring tankers, but are ready to respond to gross violations. Some states have policies to use inspections themselves as sanctions against ships that do not take pollution regulations seriously. The Dutch, for example, recently formalized the use of inspections as sanctions, noting that authorities can legally conduct an extremely thorough inspection of a vessel that can prove "very labor-

77. Kasoulides, "Paris Memorandum of Understanding," 187.

intensive and time-consuming."[78] Detentions are more standard, as well as being treaty-authorized administrative sanctions that bypass the standard legal process.[79] The new obligation to detain noncompliant tankers has elicited a less uniform response than did the right to inspect tankers. Table 5.5 shows that, from 1984 through 1990, only seven of fifteen states reporting to IMO ever detained tankers. Of these, only the United States has regularly detained more than five tankers per year. Although MOU states have detained many ships in every year since 1982, the MOU detention figures include detentions for violations of all IMO conventions, not just MARPOL. The fact that reports from MOU states to IMO also show low detention rates supports the conclusion that detentions for MARPOL violations are probably rare.

Even if infrequently, states have been enforcing oil pollution-related detentions, although none had ever done so prior to MARPOL's entry into force in 1983.[80] The United States detention rates show that violations warranting detention have been committed, although the United States has often used detention in cases involving misuse of equipment rather than its absence.[81] The infrequency of detentions by other states suggests that they have not interpreted the obligation to detain ships as strictly as the United States or, equally plausibly, that the equipment compliance rate has been high enough that such detentions have rarely been needed. Almost all states have treated the MARPOL language as if it gave them a new right rather than imposing on them a new obligation: the new provision changed enforcement behavior in that states have taken actions from which they previously had refrained, but they have done so only in rare cases.

This change in enforcement behavior cannot be attributed to exogenous increases in pressures for oil pollution enforcement. A comparison

---

78. The Netherlands, *Milieubeleidsplan voor de Scheepvaart 1991–1994* (Environmental Policy Plan for Shipping 1991–1994) (Tweede Kamer der Staten-Generaal: 22–072) (The Hague: The Netherlands Government Printing Office, 1991), 38.

79. R. L. Newbury, "Port State Control," unpublished paper, Cardiff, Wales, 1987, 6.

80. Interview, Daniel Sheehan.

81. For a description of specific cases of detention, see William P. Coughlin, "Two Ships Barred from Unloading Oil in Boston," *Boston Globe* 1 November 1990.

Table 5.5
Number of detentions

Reported to the IMO Secretariat[a]

Reporting country	1984	1985	1986	1987	1988	1989	1990
Australia					0	0	0
Bulgaria				0	0	0	1
China			0	0	0	0	0
Egypt				0	0	0	
Germany[b]		3	2	2	0	0	11
Greece					0	0	0
Hong Kong	0		0	0	0	0	0
Israel		0	0				
Italy	19	39	0				
Japan		0	0	0	0	0	0
Netherlands			0				
Norway	0	0	4	4	1	0	1
Sweden			2				
United Kingdom	0	0	0	2	47	2	3
United States	52	59	42	36	339	10	14
Total detentions reported	71	101	50	44	387	12	30
Number of reports	5	7	11	9	11	11	10
Average number per report	14	14	5	5	35	1	3
Number of ships inspected	3,602	14,610	21,879	32,332	29,957	27,444	35,243
As % of ships inspected	2.0	0.7	0.2	0.1	1.3	0.1	0.1

Reported to the MOU Secretariat[c]

All MOU member states	1984	1985	1986	1987	1988	1989	1990
Total detentions reported	476	356	307	280	295	344	441
Number of reports	14	14	14	14	14	14	14
Average number per report	34	25	22	20	21	25	32
Number of ships inspected	7,686	7,879	8,721	10,337	8,382	9,164	9,842
As % of ships inspected	6.2	4.5	3.5	2.7	3.5	3.8	4.5

Sources: Various MEPC documents from 1984 to 1992; and Secretariat of the Memorandum of Understanding on Port State Control, *Annual Report* (The Hague: The Netherlands Government Printing Office, 1984–1990).

[a]Countries include only those providing reports using IMO's mandatory reporting format. Blank spaces indicate years in which the country did not report; zeros indicate reports indicating no discrepancies.
[b]Combines reports from the Federal Republic of Germany and the German Democratic Republic for the years prior to 1990.
[c]Includes detention for violations of non-MARPOL conventions.

of the data on this behavior to the limited data available on fines suggests that the new use of detention has not represented a broader increase in commitment to environmental sanctions.[82] Average fines were not rising during the 1980s. The United Kingdom reported average fines of between $2,000 and $3,500 for 1975 through 1979, a single fine of $20,000 in 1984, and average fines of $3,241 in 1985 and $2,396 in 1989. Germany reported averages of between $407 and $1,139 in 1985 through 1990 except for 1989, when the only fine levied was for $10,372. U.S. fines for 1988 and 1989 averaged $3,000. All three states also reported imposing very few fines, all under fifteen per year, with the exception of Germany, which imposed forty in 1985 and forty-seven in 1986. Neither did any of the MOU states consistently report imposing high average fines during the 1980s. A single day of detention cost a tanker operator some $20,000 in opportunity costs, far higher than the typical fines being imposed.[83] Thus the right to detain ships succeeded in increasing sanctions, while efforts to increase fines by requiring them to be "adequate in severity" failed. Although the detention provision removed legal barriers to the use of a forceful sanction by those states with preexisting incentives to sanction, it did not make other states more likely to use it.

How do we explain the apparent success of these new provisions in inducing states to inspect ships for equipment violations and to detain violators? Tankers have received initial surveys and IOPP certificates as simple add-ons to a set of standard certification and inspection procedures required before they can trade internationally. Classification societies and flag state inspectors survey ships during construction and regularly thereafter to ensure conformance with standards for major concerns not related to pollution, such as safety. No new infrastructure has had to be created. Subsequent government equipment and certificate inspections have become common for most states for which we have information, because requiring tankers to carry certificates has significantly reduced the inspection burden, at least for governments. The

82. The data on fines come from Peet, *Operational Discharges*, Annex 15, for the periods after 1982, and prior to 1981 from Dempsey, "Compliance and Enforcement."

83. Estimates of daily costs of detention are from interviews with John Foxwell, Shell International Marine, London, 27 June 1991; and Richard Schiferli, MOU Secretariat, Rijswijk, The Netherlands, 17 July 1991.

equipment provisions have also eliminated the possibility of detecting a violation without identifying the violator.

Unlike aerial surveillance programs, equipment inspections have often been easily added to in-port inspection programs already in place to verify customs and safety compliance. No new infrastructure had to be established to extend these inspections to pollution prevention. In short, certifying tankers and detecting equipment violations has involved only marginal, not fixed, enforcement costs. Inspections have also provided greater assurance of compliance than could efforts to detect discharge violations. Even a country dedicating significant resources to the detection of discharge violations cannot take the absence of detected violations as assurance that violations are not being committed. In contrast, a certificate and equipment inspection program can be either randomized or devised to target likely violators, thereby providing enforcement agencies with vastly greater confidence that the inspections are detecting most violations. In essence, inspection for equipment violations have radically reduced the costs of being confident that enforcement resources are detecting most violations.

The MOU has increased inspections through three different processes. First, the MOU helped member states overcome the common pool resource problems inherent in enforcement, reassuring each that it could count on other states in the region to increase their port inspections. Its own efforts would not cost it directly and would provide diffuse environmental benefits to neighboring states. This reassurance was crucial to European enforcement, unlike that of the United States and Japan, which could undertake port state enforcement efforts without regional coordination since they economically dominate their respective regions. The system also helped authorities deploy enforcement resources more effectively. The MOU data base allowed inspectors to identify those ships currently in port that were most likely to pose pollution threats while avoiding the expenditure of scarce inspection resources on recently inspected ships. Additionally, the regular meetings of the parties and the annual reports focused member states' attention on the importance of enforcement. They kept enforcement high on the agendas of the maritime authorities involved.

The MOU's influence on behavior was by no means independent of MARPOL. Its success depended on provisions in MARPOL that increased states' capacity for enforcement. Until 1992 the MOU specifically excluded from its mandate cooperative enforcement against discharge violators.[84] Although the MOU states agreed to coordinate their in-port inspections, they explicitly confined themselves to inspecting for violations of provisions in IMO conventions, including MARPOL. Given that nine of the fourteen MOU states had opposed the American SBT proposals in both 1973 and 1978, it would have been surprising if these states had established SBT and IOPP certificate requirements on their own.[85] Yet these states have conducted extensive in-port inspection campaigns, including monitoring compliance with the SBT requirements. Their enforcement efforts cannot be explained by U.S. pressures to enforce MARPOL, since those were not applied. While the MOU provided the context for cooperation on enforcement, MARPOL provided the rules that the countries would enforce. Without MARPOL, how would enforcement concerns have been channeled? What would these countries have inspected? MARPOL answered these questions so that even states that opposed the equipment requirements have conducted the inspections necessary to enforce them.

Detention provisions have altered behavior because they have had the rare virtue of imposing low costs on the sanctioning agent and high costs on the violator, making their use both more credible and more potent. Detention is an administrative, not legal, penalty that involves opportunity costs rather than fines. Flag states have accepted detention because, being administrative, it did not threaten flag states' legal jurisdiction. Paradoxically, skirting the legal system has made port states more likely to use detention and flag states more likely to accept it. Detention has imposed opportunity costs that far exceed any likely fines and can force the tanker into future compliance by requiring necessary retrofits. Al-

84. Secretariat, *Annual Report 1992*. For example, cooperation among member Maritime Authorities in aerial surveillance has been exclusively outside the MOU framework.

85. Belgium, Italy, the Netherlands, and the United Kingdom fully supported the U.S. SBT proposal in 1973, although France reluctantly voted with the United States at the last moment. Only Greece, Norway, Portugal, Spain, and Sweden voted for the U.S. proposal on retrofitting SBT in 1978 (M'Gonigle and Zacher, *Pollution, Politics, and International Law,* 121 and 137).

though detention is a large enough penalty to deter a ship from committing future violations, detaining a ship until it remedies its noncompliance with equipment provisions is appropriate to the crime. Even its infrequent use by a few major oil-importing states has presented tanker owners that wish to avoid the capital investment with the alternative of taking the risk of detention or, even more costly, foregoing the oil markets in those countries deemed likely to use detention. Although the evidence shows that most states have rarely detained ships, those that have done so even infrequently control a significant share of the oil market, although not a majority. Although noncompliance with the equipment standards has been rare, states have had a system in place that could identify and respond if it did occur.

Conclusions

The two different compliance systems analyzed here have had markedly different success in influencing the behavior of actors engaged in the enforcement process. Efforts to improve the deterrence-based enforcement of the discharge standards has consistently failed among almost all countries, while the shift to equipment standards and the creation of the MOU have resulted in marked changes in behavior. Examining the two cases shows that success depended on establishing primary rules and supporting treaty provisions that placed actors within a nexus in which they had the incentives, ability, and authority to enforce treaty regulations.

Although analysts and diplomats alike hailed the 1969 amendments to OILPOL as holding the "potential for a more rigorous enforcement system," they produced little improvement in either detection or prosecution.[86] All the efforts to improve the enforcement of discharge standards faced daunting demands to remedy a wide array of possible situations in which enforcement could fail. The clean ballast provision made it more likely that evidence of a violation would be solid enough to allow for successful prosecution, but it failed to address the facts that few countries were monitoring for discharge violations, that identifying violators was difficult, and that prosecution remained in the hands of

86. M'Gonigle and Zacher, *Pollution, Politics, and International Law,* 239.

flag states with few incentives to prosecute. MARPOL's creation of explicit requirements for states to increase their detection efforts and its provision of better evidence for prosecution by flag states had few results because, although the likelihood of success was great, the resources needed to effectively detect discharge violations were enormous; lack of incentives rather than lack of evidence was the reason for poor rates of prosecution by flag states. Inherent in the original and clean ballast discharge standards were requirements for extensive surveillance programs that had low probabilities of detecting violations or identifying violators and posed difficult evidentiary obstacles to conviction.

The total discharge rule moved in the right direction by establishing a rule that dramatically decreased the monitoring demands placed on governments. It established criteria that oil-loading states could monitor in their ports days after the violation had been committed, while eliminating cases in which the perpetrators of violations could not be identified. The problem with these nominal improvements was that they failed to recognize the lack of incentives these states had to monitor such violations. Oil companies conducted such inspections, but discontinued them before the 1969 amendments took legal effect, when they could be used to identify violators. MARPOL's subsequent provision of inspection authority to port states failed because it provided legal authority to oil-loading states with few political or material incentives to look for violations.

In contrast, equipment regulations simultaneously reduced the burden of monitoring for violations, shifted much of that burden to nongovernmental actors, and left the remaining burden on developed port states with incentives to engage in inspection and detention efforts. Much of the enforcement problem was eliminated from the outset by relying on existing classification society and governmental inspection infrastructures that monitored tanker owners' behavior before violations were committed, thereby preventing them. MARPOL gave states that had preexisting incentives to detect and sanction violations the practical ability and legal authority to do so, but allowed them to focus their enforcement resources on a relatively few cases of violation. Indeed, MARPOL gave all port states the legal authority to use in-port inspections to verify compliance with both the total discharge standards and the equipment standards; the fact that states established programs to monitor compliance with the equipment standards but not to monitor compliance with the total discharge standards was due to the fact that in the latter case those with the practical ability to inspect had no incentives to do so.

Clean ballast, total discharge, and equipment standards all remain binding on MARPOL member states today. Yet enforcement of the former two provisions has increased little during precisely the same period that extensive new efforts have been made to enforce the equipment requirements. Bolstering the argument that we can attribute this difference to the treaty rules and not to exogenous factors is the fact that theory would predict less enforcement of the equipment rules since their enforcement is beneficial to all other treaty parties. Contrary to theories correlating the "privatization" of enforcement benefits with greater enforcement efforts, the United States, Japan, and the MOU countries spend far more on enforcing equipment standards—a public good that improves the global ocean environment—than on enforcing discharge standards, the benefits of which would be more "private."[87] Nongovernmental actors also would have responded quite differently if SBT requirements had been established by the United States unilaterally. Although classification societies might have encouraged most tanker owners to meet the U.S. requirements, they would have been unlikely to demand that all tankers meet an expensive standard that only the United States was requiring.

For states that lack enforcement incentives, international rules have difficulty creating them. Evidence shows that states generally fail to fulfill obligations such as those stating that they "shall" detain ships, investigate cases in which visible traces of oil are observed, and institute proceedings if sufficient evidence is available. States disinclined to enforce treaty regulations have not begun enforcing them because they were parties to a convention with such requirements. Rather, states inclined to enforce have been freed to do so when international laws that precluded strong enforcement were removed. International law appears to have greater weight in restraining enforcement activities than in encouraging it. Governments are reluctant to take enforcement actions that violate currently accepted rules and norms of sovereignty and jurisdiction.

87. Robert Axelrod and Robert O. Keohane, "Achieving Cooperation under Anarchy: Strategies and Institutions," in Kenneth Oye, ed., *Cooperation under Anarchy* (Princeton, NJ: Princeton University Press, 1986).

For instance, even though the United States had been the major proponent of SBT requirements and had the power to enforce them throughout the 1970s, it did not begin detaining tankers until after MARPOL entered into force in 1983.

Enforcement provisions must match detection and prosecution burdens with those actors that have the capacities and incentives to undertake them. MARPOL's equipment standards have increased enforcement efforts because they reduced the enforcement demands on states while placing them on states likely to carry them out. The MOU increased enforcement by working with states that were already inclined and legally authorized to enforce regulations in their ports. Detention requirements influenced the behavior of only those states that were already predisposed to sanction violations. Total discharge standards did not increase enforcement because the legal authority and practical ability to carry out the inspections necessary to enforce them were not bestowed on those likely to detect violations. Treaties must establish sets of primary rules that minimize the enforcement burdens governments face while ensuring that governments with incentives to enforce the rules have the legal authority to do so and face no legal barriers in their efforts to do so.

6

Inducing Governments to Comply: Providing Reception Facilities

For the constraints on the disposal of waste oil at sea in the 1954 International Convention for the Prevention of Pollution of the Sea by Oil (OILPOL) to have a chance of succeeding, the convention also had to ensure that tankers had an alternative means of disposal. Indeed, every international oil pollution conference since the first in 1926 has included requirements for reception facilities for oily wastes, seeing these facilities as essential to successfully eliminating ocean oil pollution. Equally consistently, however, these agreements have been ambiguous about whether governments or industry was responsible for providing these facilities, and they have created few procedures to ensure that they were provided. Not surprisingly, the response to these requirements has been mixed: reception facilities remain rare in ports in special areas and oil-exporting states, areas where they are most needed; they have been provided largely by private industry rather than governments in developed countries; and their nonexistence or inability to facilitate quick tanker discharges has been the subject of frequent complaints from the oil and tanker industries and an alleged reason for discharge violations.

The need for reception facilities has varied over time. More stringent regulations have increased the need: the 1954 rules caused reception facilities to be needed in some ports, and the total discharge limits established in the 1969 amendments to those rules significantly increased this demand. The 1973 International Convention for the Prevention of Pollution from Ships (MARPOL) reduced the total discharge limit for new tankers and prohibited all discharges in special areas, increasing the demand for reception facilities yet further. On the other hand, the adoption of new procedures and technologies has reduced the need: the use of procedures to discharge slops with subsequent cargo (the load on top or

LOT process) during the 1960s and the installation of segregated ballast tank (SBT) and crude oil washing (COW) equipment during the 1970s and 1980s all reduced the quantity of slops a tanker needs to discharge after a ballast voyage. But not all tankers use these procedures and technologies, and nontankers also generate slops. Even tankers using LOT, SBT, and COW still must use reception facilities for the slops generated on short or stormy voyages, those that cannot be combined with the next cargo, or the sludge from periodic cleanings. So the demand remains.

Reception facilities can consist of fixed storage tanks on shore or mobile facilities such as barges, moored oil tankers, railroad tankcars, or tanker trucks. The facilities need sufficient capacity—in waste storage, discharge speed, and berths—to service the traffic using that port. Particularly given a tanker's incentives to ignore the discharge regulations altogether, facilities need to be user-friendly; tankers must find it easy to locate and schedule use of the facilities without long delays or lead times and at low or no cost. Large and efficient facilities are particularly crucial in the ports of oil-exporting states, where numerous tankers returning from their ballast voyages must have a means to discharge as much as 500 tons of slops before loading new cargo if they are to meet treaty requirements.

Treaty Requirements for Reception Facilities

The 1954 convention required that governments "ensure the provision . . . of facilities adequate for the reception" of the oily bilge water and fuel tank sludge from nontankers no more than three years after the convention entered into force.[1] The resistance to any provision that would require significant government funding accounted for this weak language. The provision was purposefully ambiguous regarding who was to actually provide the facilities. Provision of the more expensive facilities for tankers at oil-loading terminals was left as a recommendation to oil comanies in the conference's final act.[2] The cost of large tanker reception

1. Article VIII, *International Convention for the Prevention of Pollution of the Sea by Oil,* 12 May 1954, 12 U.S.T. 2989, T.I.A.S. no. 4900, 327 U.N.T.S. 3, reprinted in 1 I.P.E. 332, hereinafter cited as *OILPOL 54.*

2. *Final Act of the International Conference on Pollution of the Sea by Oil,* cmnd. paper no. 9197 (London: Her Majesty's Stationery Office, 1954).

facilities, indeed, had been the source of many countries' opposition to Britain's proposal to prohibit tanker discharges throughout the oceans.[3]

The United Nations' 1956 survey had shown that the United States and Britain had relatively adequate facilities, "but if wider usage of these facilities was to be recommended, especially in loading ports, then facilities had to be improved in the Middle East and other main oil-exporting regions."[4] Even with the very weak language finally adopted, the United States reserved on it during ratification because the government did not want to assume "any financial responsibility" for building and operating such facilities.[5] The British even proposed deleting the reception facility requirement altogether because its inclusion was the basis of threats by several states not to sign.[6]

The tension between the recognition that facilities for tankers were essential to effective pollution control and the resistance of many governments to providing them led to the adoption of new language on reception facilities in the 1962 amendments to OILPOL. Rather than being relegated to the status of recommendations, the provisions in Article VIII now addressed facilities in oil-loading ports as well as non-tanker ports. At the same time, the United States succeeded in weakening the requirement so that governments merely needed to "take all appropriate steps to promote the provision of facilities," and the date for ensuring their provision was completely eliminated.[7]

Recognizing cost as an important source of existing and future noncompliance, Greece proposed establishing an international fund to pay

3. Sonia Zaide Pritchard, *Oil Pollution Control* (London: Croom Helm, 1987), 102.

4. Pritchard, *Oil Pollution Control,* 81.

5. 12 U.S.T. 3024 (1961), cited in Charles Odidi Okidi, *Regional Control of Ocean Pollution: Legal and Institutional Problems and Prospects* (Alphen aan den Rijn, The Netherlands: Sijthoff and Noordhoff, 1978), 33; and see United States Senate, "Message from the President," *Amendments of the International Convention for the Prevention of Pollution of the Sea by Oil, 1954* (88th Congress, 1st session) (Washington, DC: GPO, 1963), 19.

6. Pritchard, *Oil Pollution Control,* 128.

7. Article VIII, *International Convention for the Prevention of Pollution of the Sea by Oil, as Amended in 1962,* 11 April 1962, 600 U.N.T.S. 332, reprinted in 1 I.P.E. 346, hereinafter cited as *OILPOL 54/62.*

for needed reception facilities, but the proposal received little support.[8] The ongoing debate over whether governments or industry was responsible for providing reception facilities was starkly evident both in the ambiguous amendment language calling on governments to promote the provision of facilities and in the resolution in the conference's final act that reception facilities at oil-loading terminals, "where they still do not exist, should now be provided as a matter of urgency by those organizations which have it within their means to provide them or to secure or promote their provision."[9] At least with respect to facilities for oil tankers, government expected industry to pick up the tab.

The growing pressures for environmental improvement that led the 1973 conference to develop a broader and more stringent regulatory structure also created a climate more conducive to strengthening the reception facility provisions. Although the most recent survey of facilities had been published in 1964, delegates realized that the underlying goal of the 1969 amendments and MARPOL "that wastes should be retained on board for eventual discharge into shore reception facilities" could clearly not be achieved given the number and quality of existing facilities.[10] Indeed, the United States, having changed its tune since 1962, saw reception facility requirements as "an essential element of [MARPOL] because without the reception facilities it becomes meaningless to require retention of residues on board ship."[11] MARPOL required that all countries "ensure the provision" of reception facilities in all ports "where ships have oily residues to discharge," within one year of the convention's entry into force.[12] Substantive and bureaucratic obstacles to MARPOL's ratification kept the provision from taking effect until October 1983.

8. Pritchard, *Oil Pollution Control*, 129.

9. See *OILPOL 54/62* and IMCO, *Resolutions Adopted, 1962*.

10. IMO/UNDP, *Proceedings of IMO/UNDP International Seminar on Reception Facilities for Wastes* (London: IMO/UNDP, 1984), iv.

11. Rear Admiral H. H. Bell, "Statement," in United States Senate, Committee on Foreign Relations, *Hearings on Protocol of 1978 Relating to the International Convention for the Prevention of Pollution from Ships, with Annexes and Protocols: June 12, 1980* (96th Congress, 2nd session) (Washington, DC: GPO, 1980), 7.

12. Annex I, Regulation 12, par. 1, *International Convention for the Prevention of Pollution from Ships*, 2 November 1973, reprinted in 12 I.L.M. 1319 (1973),

Reception facilities were especially crucial in the environmentally sensitive "special areas" in which MARPOL prohibited all oil discharges, requiring all slops to be discharged at shore facilities. The drafters of MARPOL sought to ensure that facility provision in these areas did not fall victim to the delays in ratification that had plagued earlier agreements. Therefore, for facilities in special areas bordered predominantly by developed states (the Baltic, Black, and Mediterranean Seas), MARPOL set a compliance deadline of 1 January 1977, after which the stricter discharge standards would become operative.[13] Recognizing the greater difficulties facing the developing states bordering the Red Sea and Persian Gulf regions, MARPOL required facilities only "as soon as possible."[14] However, the complete discharge prohibition would not take effect until these countries notified the Intergovernmental Maritime Consultative Organization (IMCO) that sufficient and adequate facilities had been provided. The same requirements applied to the Gulf of Aden when it was designated as a special area in 1987. To address the obstacles to compliance among developing states, MARPOL included the construction of reception facilities on the list of technical assistance projects that it urged developed countries to help finance.[15] The 1978 MARPOL Protocol Conference left all of these provisions unaltered.

What processes could have been expected to lead states to meet these requirements? The efforts of the International Maritime Organization (IMO) and member nations to promote the provision of reception facilities can be grouped into four categories: collecting and disseminating information on available reception facilities,[16] conducting regional studies to identify problem areas,[17] examining measures and conducting

---

2 I.P.E. 552; and *Protocol of 1978 Relating to the International Convention for the Prevention of Pollution from Ships,* 17 February 1978, reprinted in 17 I.L.M. 1546 (1978), 19 I.P.E. 9451, hereinafter cited as *MARPOL 73/78*.

13. Annex I, Regulation 10(2), *MARPOL 73/78*.

14. Regulation 10(7), Annex I, *MARPOL 73/78*.

15. Article 17, *MARPOL 73/78*.

16. See chapter 4's description of the various surveys of facilities conducted and published by IMCO/IMO in 1956, 1964, 1973/76, 1980/84, and 1990. See also MEPC 26/Circ. 135 (22 March 1988).

17. See J. Wonham, "Summary of Initiatives by IMO and Other Related Organizations Aimed at Focusing Attention on the Reception Facility Require-

seminars on how to reduce the costs of providing and operating facilities,[18] and urging developing nations to request and developed nations to provide funds for the building and operation of facilities.[19]

Besides publishing lists of available reception facilities, IMO bodies have frequently passed resolutions noting the inadequacy of reception facilities, especially in developing states, and urging states to fulfill their MARPOL obligations.[20] Italy and the Commission of the European Communities sponsored a study on Mediterranean reception facilities, and the Regional Organization for the Protection of the Marine Environment studied reception facility problems in the Persian Gulf.[21] Studies have evaluated the use of old tankers as inexpensive floating facilities and ways to make waste oil recycling more profitable.[22] IMO's frequent discussions since the original Greek proposal in 1962 regarding developed countries' financing developing countries' reception facilities have never engendered sufficient support to put these ideas into practice.

Assessing the Provision of Reception Facilities

Are governments actually ensuring that their ports have adequate reception facilities? If so, to what degree are the OILPOL and MARPOL

---

ments of MARPOL 73/78," in IMO/UNDP, *Proceedings,* 200–202; MEPC/Circ. 102 (20 July 1982); and MEPC/Circ. 118 (22 December 1983).

18. See, for example, "International Maritime Organization in 1985 and 1986," in Elisabeth Mann Borgese and Norton Ginsburg, eds., *Ocean Yearbook 6* (Chicago: University of Chicago Press, 1986), 413.

19. See, for example, MEPC/Circ. 210 (14 September 1988); MEPC 27/5/3 (7 February 1989); and P. G. Sadler and J. King, "Study on Mechanisms for the Financing of Facilities in Ports for the Reception of Wastes from Ships," MEPC 30/Inf. 32 (12 October 1990) (Cardiff, Wales: University College of Wales, Cardiff, 1990).

20. In 1985 the IMO Assembly noted that "in many areas of the world [facilities] are still inadequate" ("International Maritime Organization in 1985 and 1986," 413).

21. IMO/UNDP, *Proceedings;* and Sadler and King, "Study on Mechanisms," 1990.

22. See S. Fukuoka, "Presentation by Nippon Kokan (NKK) on Floating Reception Facilities," in IMO/UNDP, *Proceedings;* and MEPC 27/Inf. 2 (1989).

requirements themselves, and the associated IMCO/IMO procedures, responsible for this progress? The short answers are that, in oil-exporting and developing states near special areas, reception facilities remain either nonexistent or woefully inadequate to handle the volume of tanker traffic they receive. Developed states now have many more facilities in more of their ports than they did in the 1950s and 1960s. The progress made, however, appears to have been a slow, gradual response to domestic environmental pressures and the recurring discourse on reception facilities in various IMO fora rather than a response to specific treaty requirements and timetables.

Table 6.1 summarizes the responses to IMO surveys on reception facilities that were published in 1956, 1963, 1973/76, 1980/84, and 1990 and an IMO-commissioned study conducted in 1990. The surveys differed considerably in terms of which and how many countries reported. The reports varied—across countries and for a single country over time—as to how many ports they provided information on, how they defined reception facilities, and the extent to which they identified ports lacking reception facilities. Some responses did not even specify how many ports they were providing information on, merely specifying that none or all of their ports had reception facilities. The table summarizes this complex and disparate data by combining all ports with any form of reception facility, whether large fixed facilities, inadequately-sized facilities, ad hoc arrangements for a tanker truck to receive slops, or a facility that accepts only some types of oil waste.

The results of the IMO surveys from 1956 to 1990 showed that a relatively constant number of countries reported on a growing number of ports and a steadily increasing number of facilities (major ports often have more than one facility). The number of ports without facilities remained relatively constant (between twenty and forty) until it jumped to 104 in 1990. These trends reflect a growing number of ports with reception facilities and a generally declining fraction of ports reported without them.

Reception Facilities and the Level of Countries' Development

This apparent progress at the aggregate level obscures considerable variation when the data are broken down. Indeed, combining all countries' reports introduces numerous problems. For example, the United States

Table 6.1
Surveys on reception facilities in ports

	1956 IMO	1964 IMO	1973/6 IMO	1980/4 IMO	1990 IMO	1990 Sadler and King
All reporting countries						
Number of countries reporting	40	31	27	40	37	129
Number of ports	162	189	353	508	993	478
Number of reception facilities	402	475	654	847	1765	na
Number of ports without reception facilities	37	31	37	22	104	151
% of ports without reception facilities	22.8	16.4	10.5	4.3	10.5	31.6
United States						
Number of ports	14	14	55	39	325	49
Number of reception facilities	217	232	266	59	948	na
Number of ports without reception facilities	0	0	0	0	36	4
% of ports without reception facilities	0.0	0.0	0.0	0.0	11.1	8.2
OECD (non-U.S.) countries						
Number of countries reporting	19	18	17	18	18	22
Number of ports	121	149	260	390	575	212
Number of reception facilities	169	217	345	635	739	na
Number of ports without reception facilities	27	22	36	14	48	38
% of ports without reception facilities	22.3	14.8	13.8	3.6	8.3	17.9
OPEC countries						
Number of countries reporting	3	1	1	2	1	13
Number of ports	9	0	4	2	1	76
Number of reception facilities	8	0	4	1	0	na
Number of ports without reception facilities	1	0	0	1	1	47
% of ports without reception facilities	11.1	na	0.0	50.0	100.0	61.8
Non-OPEC/non-OECD countries						
Number of countries reporting	17	11	8	19	17	93
Number of ports	18	26	34	77	92	141
Number of reception facilities	8	26	39	152	78	na
Number of ports without reception facilities	9	9	1	7	19	62
% of ports without reception facilities	50.0	34.6	2.9	9.1	20.7	44.0

Sources: United Nations Secretariat, *Pollution of the Sea by Oil* (New York: United Nations, 1956); IMCO, *Pollution of the Sea by Oil* (London: IMCO, 1964); IMO, *Facilities in Ports for the Reception of Oil Residues* (London: IMO, 1973, 1976, 1980, 1984); MEPC Circ. 234 (13 August 1990); and P. G. Sadler and J. King, *Study on Mechanisms for the Financing of Facilities in Ports for the Reception of Wastes from Ships* (Cardiff, Wales: March 1990).

did not report a single port as without a reception facility before 1990 and then reported thirty-six. It also reported on almost five times as many ports and reception facilities in 1990 as in any previous year. Other countries have also altered their reporting criteria: excluding the United States, members of the Organization for Economic Cooperation and Development (OECD) increased the number of ports reported from 121 in 1956 to 575 in 1990. Norway reported on 9 ports in 1963, 4 ports in 1973, 22 ports in 1980, and 96 ports in 1990, but has never identified ports lacking reception facilities. In general, the reports have probably underestimated the share of ports without reception facilities, since states without facilities may not have reported and even reporting states have often not identified ports that lacked reception facilities. To remedy some of these problems, the table uses four development categories—the United States, non-United States developed (OECD) countries, members of the Organization of Petroleum-Exporting Countries (OPEC), and developing (non-OECD, non-OPEC) countries.

When the results of the surveys are disaggregated, it becomes clear that almost all the increases in reception facilities occurred among developed countries. Most OECD countries have regularly reported, providing initial confidence in the data. Of the 1,363 facilities that were newly reported between 1956 and 1990, 1,301 were in the United States and other OECD states. In these countries the number of reception facilities, the number of ports with reception facilities, and the percentage of ports with reception facilities have all increased steadily since 1956. The OECD states posted a marked increase between the 1973 and 1980 surveys. While some of the increase certainly stemmed from greater detail in reporting (it is unlikely that 130 new ports were established between 1973 and 1980), it also seems unlikely that 290 reception facilities existed in these countries but were not included in their 1973 reports. Most likely new facilities were installed in some cases, existing sludge disposal services were relabeled as reception facilities, and new disposal services were established. Many of the facilities have been provided by private companies, although the reporting has not been systematic enough to allow us to evaluate how many.

Independent data tend to confirm the IMO survey data from developed states. In 1983, 1985, and 1990, the International Chamber of Shipping

Table 6.2
International Chamber of Shipping (ICS) surveys on reception facilities in port

	1983	1985	1990
All reporting countries			
Number of countries reporting	78	70	51
Number of ports without reception facilities	197	215	114
United States			
Number of ports without reception facilities	10	26	12
OECD (non-U.S.) countries			
Number of countries reporting	13	15	10
Number of ports without reception facilities	63	64	31
OPEC countries			
Number of countries reporting	12	12	8
Number of ports without reception facilities	46	41	20
Non-OPEC/non-OECD countries			
Number of countries reporting	52	42	32
Number of ports without reception facilities	78	84	51

Sources: MEPC 19/5/2 (21 October 1983); MEPC 22/8/2 (8 October 1985); and MEPC 30/Inf. 30 (15 October 1990).

(ICS) asked tanker captains to report on ports where they could not locate reception facilities or found inadequate facilities. Although captains may not know about existing facilities due to poor communications with tanker agents and port authorities, they have greater incentives to identify ports lacking facilities than national governments. Table 6.2, depicting tanker captains' reports on ports without reception facilities, shows that this independent source of data provided figures comparable to those of the self-reports by the United States and OECD states. The one systematic effort to identify the availability and adequacy of facilities worldwide, which was sponsored in 1990 by IMO and conducted by P. G. Sadler and J. King, also showed comparable figures for the numbers of ports without reception facilities (see table 6.1).

The availability of facilities does not always indicate that they are accessible, cheap, or easy to use, however. The ICS surveys, the Sadler and King study, and irregular reports to IMO indicate that even developed states' reception facilities frequently fail to meet any reasonable criteria for adequacy.[23] Overall, however, developed states have significantly improved their provision of reception facilities. Germany has recently even conducted a pilot program making reception facilities available free of charge.[24] Unfortunately, similar improvements have not been made in OPEC or other non-OECD countries.

Reports to IMO from developing states have provided data on an increasing number of ports and have shown growth in the number of reception facilities but considerable variance in the number and percentage of ports without facilities. Even these minimal conclusions remain suspect, however, because most developing countries have not furnished information. Even if the IMO statistics are accurate for the countries reporting, they omit far too many countries. The ICS and Sadler and King studies provide more comprehensive data that show many more ports without facilities. Of the 141 ports in the Sadler and King study, 44 percent had no reception facilities, and a considerably greater number had inadequate ones. In contrast, this study identified only 8 percent of United States ports and 18 percent of other OECD ports as without reception facilities.

Especially disturbing is the almost complete absence of reception facilities in OPEC countries. Never have more than three of the thirteen OPEC members responded to IMO surveys. Since 1956 the responses received have identified four or fewer facilities. Although it is in the ports of these oil-loading states that tankers have the greatest need for reception facilities, both the ICS and Sadler and King data show that there are more than forty ports without any facilities and many more with inadequate ones. Sadler and King estimate that over 60 percent lack facilities of any kind.

23. See, for example, MEPC/Circ. 135 (18 March 1985); MEPC/Circ. 135/Add. 1 (22 March 1988); and MEPC/Circ. 135/Add. 2 (8 August 1988) as well as the ICS surveys and the Sadler and King study.

24. R. P. Barston, "Conference Reports: Implementing IMO Agreements," *Marine Policy* 13 (July 1989), 267.

In short, developed states are considerably more likely than developing states to have reception facilities in their ports. Whether through the provision of direct government funding, the application of pressures and incentives to private companies, or other methods, these countries have made it quite likely that a tanker seeking to use facilities can do so. In contrast, developing states and OPEC states have done considerably less to ensure that such facilities are available.

Reception Facilities in Special Areas
The most notable change in reception facility requirements was made when MARPOL replaced the earlier language stating that countries should "promote provision" of reception facilities with the requirement that countries bordering special areas "ensure provision" of facilities more quickly than other countries. Mediterranean, Baltic, and Black Sea states faced a deadline of 1 January 1977, and states bordering other special areas needed to install facilities "as soon as possible." If the rules were effective, reception facilities in these regions should have increased more rapidly after MARPOL was signed and more rapidly than in other regions. However, neither of these has been the case.

As table 6.3 shows, the number of special area ports with reception facilities and the total number of facilities have increased steadily since 1956, while the percentage of ports without reception facilities has declined. Although the number of reported facilities almost tripled between 1973 and 1980, this was due largely to the fact that the 1980 data included much greater detail by the Baltic Sea states. The IMO data have also proved unreliable since only about one-third of countries bordering special areas have reported to IMO. Sadler and King's more comprehensive 1990 data show that, when all forty-one such states are accounted for, the percentage of ports lacking reception facilities rises from the IMO figure of 6 percent to 31 percent (sixty-three ports). ICS data reinforce these figures, with captains reporting that between thirty and seventy ports had no facilities.

Sadler and King estimated that ports in special areas are no more likely (31 versus 32 percent) to have reception facilities than ports elsewhere. The Baltic Sea, surrounded by five developed states, the former Soviet Union, and Poland, is the only special area that has met the requirement to provide adequate reception facilities.[25] The Red Sea, Persian Gulf, and Gulf of Aden states have failed to comply with MARPOL's facility requirements and, as a consequence, the stricter special area standards have yet to be implemented.[26] If such implementation had also been contingent on compliance with reception facility requirements for the Mediterranean and Black Seas, these special areas would also remain unimplemented. Within these regions, in 1990 six MARPOL states had failed to provide any reception facilities and another had only limited facilities. Evidence from ICS surveys and Sadler and King's study—and the indirect evidence that if facilities were available these countries would have incentives to inform IMO—suggest that many ports bordering special areas still lacked adequate facilities nineteen years after the treaty was signed and eight years after its entry into force.

Reception facility provision in the Mediterranean has received the most attention, because major oil tanker routes run through its center and several developed states concerned about oil pollution border it. Yet even there the record is poor. An IMCO/UNEP study in 1977 (the year by which MARPOL required facilities in all Mediterranean ports) concluded that, of nineteen crude oil-loading terminals, fewer than half had facilities, some of which were inadequate.[27] A 1981 study found only sixteen of fifty-two ports (thirty-one percent) in eleven Mediterranean countries had adequate reception facilities. Nor were inadequate facilities limited to developing countries, with some found in France, Greece, Italy and Spain.[28] Despite the Mediterranean action plan's creation in 1975, and the increased attention on pollution that it reflected and reinforced, the

25. MEPC 29/21/3 (15 January 1990).

26. See L. Andren and D. Liu, "Environmentally Sensitive Areas and Special Areas under MARPOL 73/78," in *IMAS 90: Marine Technology and the Environment* (London: Institute of Marine Engineers, 1990); and Sadler and King, "Study on Mechanisms."

27. C. L. Montfort, "Presentation of the Outcome of the IMO/UNEP Feasibility Study on Reception Facilities in a Special Area—Mediterranean," in IMO/UNDP, *Proceedings*, 228. Original document in MEPC XI/16 (25 April 1979).

28. C. Placci, "Report on the European Economic Commission/ENI-Snamprogetti (Italy) Technical and Economic Study on the Installation of Ballast Water Treatment Facilities at Mediterranean Ports and Oil Terminals," in IMO/UNDP, *Proceedings*, 292–302. See also Montfort, "Presentation of the Outcome," 250.

Table 6.3
Reception facilities in ports bordering special areas

	1956 UN	1964 IMCO	1973/6 IMCO
Special area countries			
Number of countries reporting	15	11	12
Number of ports	43	79	139
Number of reception facilities	30	70	153
Number of ports without reception facilities	25	20	25
% of ports without reception facilities	58.1	25.3	18.0
Non–special area countries			
Number of countries reporting	25	20	15
Number of ports	119	110	214
Number of reception facilities	372	405	501
Number of ports without reception facilities	12	11	12
% of ports without reception facilities	10.1	10.0	5.6
Mediterranean countries			
Number of countries reporting	9	5	5
Number of ports	32	50	52
Number of reception facilities	20	32	66
Number of ports without reception facilities	22	18	3
% of ports without reception facilities	68.8	36.0	5.8

Sources: United Nations Secretariat, *Pollution of the Sea by Oil* (New York: United Nations, 1956); IMCO, *Pollution of the Sea by Oil* (London: IMCO, 1964); IMO, *Facilities in Ports for the Reception of Oil Residues* (London: IMO, 1973, 1976, 1980, 1984); MEPC/Circ. 234 (13 August 1990); P. G. Sadler and J. King, *Study on Mechanisms for the Financing of Facilities in Ports for the Reception of Wastes from Ships* (Cardiff, Wales: March 1990); MEPC XI/14 (25 April 1979); C. Placci, "Report on the European Economic Commission/ENI–Snamprogetti (Italy) Technical and Economic Study on the Installation of Ballast Water Treatment Facilities at Mediterranean Ports and Oil Terminals" in *Proceedings of IMO/UNDP International Seminar on Reception Facilities for Wastes* (London: IMO/UNDP, 1984); and MEPC 26 INF. 23 (5 August 1988).

Table 6.3
(continued)

1979 IMCO/UNEP	1980/4 IMCO	1984 IMO/UNEP	1988 IMO/UNEP	1990 IMO	1990 Sadler and King
	16			16	41
	246			304	204
	413			381	na
	6			18	63
	2.4			5.9	30.9
	24			21	88
	262			689	274
	434			1,384	na
	16			86	88
	6.1			12.5	32.1
16	8	11	7	8	17
79	67	59	46	97	103
na	84	na	na	145	na
16	6	16	10	17	29
20.3	9.0	27.1	21.7	17.5	28.2

1983 ICS survey showed "many ports and terminals in the Mediterranean Sea were either without facilities for oil residues or lacked adequate facilities for the needs of ships using those ports and terminals."[29] In 1990, seven of sixteen Mediterranean states lacked any facilities. In light of such problems, France decided in 1990 not to enforce MARPOL's discharge standards against French tankers that had loaded oil in Mediterranean ports lacking reception facilities.[30] In short, all available evidence suggests that special area countries have provided facilities no faster than countries outside these areas.

Analysis

To summarize the information already presented, compliance with OILPOL and MARPOL's reception facility requirements has been mixed. Provision of adequate reception facilities correlates far better with a country's level of economic development than with its being in a special area or having passed a MARPOL deadline. Developed states have considerably more reception facilities than they did in 1956, and only a relatively small, though not insignificant, fraction of ports remain without adequate facilities. Some developing states have also made more reception facilities available, but far more states have not addressed the problem at all and a majority of ports have no facilities or sorely inadequate ones. Especially disturbing is the lack of progress in oil-loading ports and ports in special areas, where reception facilities are most needed. OPEC states have particularly poor records in ensuring that their oil-loading ports have facilities for tankers returning from ballast voyages. Ports in special areas are no more likely than ports in other regions of the world to have the necessary facilities. The states in special areas

29. M. D. Squires, "Results of an Enquiry by the International Chamber of Shipping (ICS) on the Adequacy of Reception Facilities for Oil Residues from Ships," in IMO/UNDP, *Proceedings*, 70. On the Mediterranean Action Plan, see Peter M. Haas, "Do Regimes Matter? Epistemic Communities and Mediterranean Pollution Control," *International Organization* 43 (Summer 1989); and Peter M. Haas, *Saving the Mediterranean: the Politics of International Environmental Cooperation* (New York: Columbia University Press, 1990).

30. Sadler and King, "Study on Mechanisms," 6. See also MEPC 19/5/1 (7 October 1983).

that have facilities are predominantly developed states. In none of the three special areas covered by the January 1977 deadline (the Baltic, Black, and Mediterranean Seas) did the deadline noticeably influence behavior. The other special areas (the Red Sea, Persian Gulf, and Gulf of Aden) were urged to provide facilities "as soon as possible," but these special areas remain unimplemented twenty years after MARPOL was signed precisely because reception facilities are not available.

How do we explain this record of compliance? Where additional reception facilities have been provided, to what extent has their provision been in response to OILPOL's requirements, MARPOL's later refinements, or any associated treaty processes? Where facilities remain nonexistent or inadequate, what explains the noncompliance?

Among developed states, reception facility provision increased most between the mid-1970s and the early 1980s. Without precise data on when facilities were constructed, the exact causes of their provision cannot be identified. Construction during this period could reflect developed states bordering the Baltic and Mediterranean complying with the January 1977 deadline, other developed states meeting the general deadline of 1984 one year after MARPOL took effect, or a less direct response to growing environmental pressures at home, economic factors, and ongoing efforts at IMO by the United States and the oil and shipping industries to raise the salience of the reception facility issue internationally. All these factors have likely played roles.

Developed states have more resources and greater domestic political incentives to protect the environment than do developing states. The larger number of reception facilities in the former undoubtedly reflects both these factors. However, governments would have been unlikely to express their concern in greater provision of reception facilities in the absence of MARPOL. MARPOL provided international legitimacy to growing domestic pressures in these countries for marine protection in the wake of the numerous oil spills in the 1970s and early 1980s. More important, MARPOL's regulations provided a readily available policy focus for these pressures, allowing governments to channel domestic pressures into programs to provide facilities directly or to require industry to provide them.

The United States provides a case study of how treaty requirements influenced provision of reception facilities, as well as the tension between government concern that facilities be provided and government reluctance to provide them. The United States originally had opposed international reception facility requirements that placed financial responsibility for providing facilities on the federal government. It spent few resources to meet OILPOL's requirements: between 1956 and 1973 the United States reported fewer than fifty new facilities, despite providing information on more than forty additional ports. No new facilities were built at all in the mid-1960s. By 1973 the increasingly environmental United States promoted the more stringent facility requirements. The subsequent increase in American reception facilities, however, has been largely the result of the rising price of oil during the 1970s, which made reclamation viable, and private oil reclamation companies have provided almost all the new facilities.[31] The Coast Guard explicitly made port authorities and oil terminal operators responsible for providing reception facilities, and it also prohibited any tanker from entering a port that did not have a Coast Guard-issued certificate of adequacy.[32]

These stringent rules and strong incentives were directly related to the MARPOL ratification process. The Coast Guard developed these rules in anticipation of congressional pressures to meet the terms of the treaty without requiring increased funding to do so.[33] American governmental concern over oil pollution extended to ensuring that port authorities provided reception facilities, but not to funding them. It seems unlikely, however, that oil pollution concerns would have found policy expression in such strong domestic regulations in the absence of MARPOL's requirements.

Reinforcing these pressures was a sense, fostered in IMO, that as tankers began to comply with MARPOL's equipment standards, governments had to ensure that their ports could accept the resultant waste oils as the other end of the bargain. IMO's Marine Environment Protection Committee (MEPC) has "acknowledged many times that full compliance

31. Okidi, *Regional Control of Ocean Pollution*, 33.

32. See Ellis H. Davison, II, "Legislative/Administrative Mechanisms in the United States for Ensuring that Adequate Reception Facilities Are Provided in Ports," in IMO/UNDP, *Proceedings,* 47.

33. Davison, "Legislative/Administrative Mechanisms."

by ships with all MARPOL discharge requirements is contingent upon the availability of adequate facilities in ports."[34] The French urged that discharge provisions be waived for ships going to ports without reception facilities. Such discussions highlight the link between reception facilities and clean seas. Tanker owners complain that, while they have spent billions on compliance with equipment standards, governments have failed to meet what tanker owners view as a government responsibility to provide reception facilities, thereby forcing tankers to violate discharge regulations.[35]

For developed states, treaty requirements and discussions within IMO combined with economic factors during the 1970s to focus domestic policy forces on ensuring that ports had reception facilities. Financially capable and environmentally concerned developed states in general responded to these forces. In contrast, these same treaty requirements, discussions, and economic factors have had little if any impact in developing and oil-exporting states. Both financial incapacity and an absence of political motivations are at work. Developing states, and especially OPEC states, have placed environmental protection considerably lower on the list of domestic priorities than many other states have. No domestic political constituency comparable to that in the United States exists to push for environmental protection. Only three members of OPEC have signed MARPOL, although an additional seven of the thirteen members are bound by OILPOL's facility provisions. Among these states, the failure to provide facilities is clearly due to a lack of political will. As the World Bank noted in 1991, "the cost of providing reception facilities in ports should be absorbed into general port costs" rather than being paid through government expenditures.[36] In cases like those of Kuwait, Bahrain, Iraq, Libya, and Oman, which have limited reception facilities or none, major oil revenues make claims of an incapacity to comply suspect.

34. MEPC 27/5/3 (7 February 1989).

35. See, for example, MEPC 19/5/1 (7 October 1983); MEPC 27/5 (17 January 1989); MEPC 27/5/3 (7 February 1989); and MEPC 27/5/4 (15 February 1989).

36. Regional Marine Pollution Emergency Response Centre for the Mediterranean Sea, *Review of the Current Situation Concerning Reception Facilities for Ship-generated Wastes in Mediterranean Ports* REMPEC/WG.3/INF.5 (7 November 1991).

A more plausible explanation is the incidence of costs: costs of providing reception facilities fall precisely on countries with few incentives to provide them.

In addition, non-OPEC developing states face both technical and financial obstacles to providing facilities.[37] One 1984 estimate put the cost of an average facility at $3.125 million, with $145 million needed to bring reception facilities in fifteen Mediterranean states up to par.[38] Another source estimated that average costs would range from $600,000 for a facility in a small port to $21 million in an oil-loading terminal, with $133 million needed for thirteen Mediterranean states.[39] A 1992 working group of the United Nations Conference on Environment and Development (UNCED) estimated that installing oily waste reception facilities in developing countries would cost $560 million for the period from 1993 to 2000.[40]

Whether noncompliance arose from an absence of capacity or of incentives, financial mechanisms could have overcome the problem, but IMO has never established a program to finance facility costs for developing countries. As attested to by both MARPOL's ambiguity over whether industry or government is to provide facilities and the U.S. experience, the governments of developed states have been reluctant to fund facilities in their own ports, let alone elsewhere. In 1962 Greek and French proposals to fund facility construction through an international tax on oil imports were rejected.[41] In 1977 MEPC noted the financing problem but did not develop a program to address it.[42] In 1980 the Secretary-General of IMCO offered Tunisia technical advice, "but re-

37. MEPC 27/5/3 (7 February 1989).

38. Montfort, "Presentation of the Outcome," 249.

39. Placci, "Report on the European Economic Commission Technical and Economic Study," 296–302.

40. Preparatory Committee for the United Nations Conference on Environment and Development, *Protection of Oceans, All Kinds of Seas Including Enclosed and Semi-enclosed Seas, Coastal Areas and the Protection, Rational Use and Development of Their Living Resources* U.N. Doc. A/Conf. 151/PC/100/Add. 21 (New York: United Nations, 1991).

41. Pritchard, *Oil Pollution Control,* 129.

42. MEPC VIII/17 (14 December 1977).

minded delegations that the financial burden for the installation of facilities should be borne by the countries concerned."[43] Even thirty years after OILPOL first required reception facilities, European states were only "looking into the possibility of eventual financial support for the development of reception facilities."[44] There is no evidence of reception facilities having been financed by donations from developed countries, either with or without IMO assistance. IMO has financed studies to identify the need for and estimate the costs of providing reception facilities in developing countries, but has never followed through by arranging for funding of these projects. While collective financing of developing states' compliance by developed states has become an increasingly frequent component of international environmental accords, it seems unlikely to become a mechanism for increasing the provision of reception facilities any time soon.

Although positive incentives for compliance have been rejected, no sanctions by government or industry have been "provided where a state fails to make adequate facilities available."[45] At the domestic level, national governments can use domestic law to clarify MARPOL's language and penalize port authorities for not providing facilities as the French, British, and Americans have done.[46] The Independent Tanker Owners' Association (INTERTANKO) attributes the "remarkably low activity, in the efforts to comply" with MARPOL's reception facility provisions to the inability of shipping interests "to fine or influence authorities to penalize ports that neglect their obligations."[47] Although

43. MEPC XII/14 (3 January 1980), 26.

44. MEPC 20/19 (24 September 1984), 19.

45. William T. Burke, Richard Legatski, and William W. Woodhead, *National and International Law Enforcement in the Ocean* (Seattle: University of Washington Press, 1975), 207.

46. See D. Powles and C. L. Montfort in D. Powles, "Charter Party Provisions Applicable to Tankers Engaged in Trades Where Deballasting Facilities Are Not Available," in IMO/UNDP, *Proceedings,* 1984, 156–157; P. Hambling, "Summary of the Approach Taken in the United Kingdom to the Implementation of MARPOL 73/78 Reception Facility Requirements," in IMO/UNDP, *Proceedings;* and Davison, "Legislative/Administrative Mechanisms."

47. MEPC 27/5/4 (15 February 1989).

IMO has passed general exhortations calling on all countries to improve facility provision, no efforts have ever been made to publicly identify and pressure noncompliant countries—for example, via a "blacklist."

Finally, reception facilities have not been provided in part because of MARPOL's ambiguity regarding responsibility for their provision. Governments need not provide facilities, but only "ensure" their provision. However, "the means whereby Governments will meet their commitments differ from one country to another since they depend on the type of the direct relationship between such Governments and the port authorities."[48] At the international level, MARPOL's language remains unclear as to who should pay.[49] Oil and shipping interests argue that provision of reception facilities should be a government responsibility, all the more so in light of the billions of dollars they have spent complying with MARPOL's equipment standards.[50] In contrast, governments argue that the "polluter pays principle" dictates that those benefiting from the polluting activity should pay the costs of eliminating it.[51] This argument proves convenient both for oil-loading states, which can argue that oil companies should provide their own terminals with facilities, and for developed states, which can argue that they should not subsidize polluting industries either at home or by financing facilities in developing states.

48. C. L. Montfort quoted in Powles, "Charter Party Provisions," 157.

49. "In Defence of the Oceans: Interview with William A. O'Neil," Horizon (Smit International) 1 (November 1990), 4.

50. See T. A. Meyer, "Alternative Tanker Operating Procedures and Some Commercial Aspects on Trades Where Reception Facilities Are Inadequate," in IMO/UNDP, *Proceedings*, 126; and Sadler and King, "Study on Mechanisms," 5. For similar arguments as far back as 1926, see Pritchard, *Oil Pollution Control*, 4–5. In the early 1980s many tanker owners began inserting clauses in charter agreements "to shift the burden of providing reception facilities from the owner to the charterer [usually an oil company] and require him to bear the costs of retaining oily wastes on board where no facilities are available and of ultimately discharging them" (Powles, "Charter Party Provisions," 146).

51. See United Kingdom, Department of the Environment, *Oil Pollution of the Sea* (Pollution Paper no. 20) (London, England: Her Majesty's Stationery Office, 1983), 23; and R. Michael M'Gonigle and Mark W. Zacher, *Pollution, Politics, and International Law: Tankers at Sea* (Berkeley: University of California Press, 1979), 116.

Conclusions

The major obstacle to provision of reception facilities appears not to be a lack of capacity, but rather a lack of incentives. The same rules have elicited quite different responses depending on whether a given state had incentives to provide reception facilities. While compliance with MARPOL's reception facility clauses correlates well with countries' level of development, the analysis has shown that the level of development is largely a proxy for the incentives to encourage reception facility provision. All states have the legal authority to provide reception facilities. Developed states and oil-exporting states also have the necessary resources. However, only the former have done so, driven largely by domestic political forces that provide incentives independent of the treaty.

Although whether and when developed states have taken action to prevent ocean oil pollution has depended on the domestic political incentives they faced, MARPOL has largely defined the menu of policies these countries have chosen from to satisfy these domestic political pressures. When government officials have felt the need to "do something" about oil pollution, promulgating requirements for ports to provide reception facilities or instigating programs to increase their use have often been placed on the agenda because they are required by the convention. The claims of oil and shipping interests that their own compliance with equipment rules deserves reciprocity and their charges that absent or inadequate facilities are now the primary reason for illegal discharges can, if not move governments to act, at least channel government pollution prevention efforts in certain directions. The shifts over the years to more stringent schedules and requirements have not been accompanied by sanctions or financial inducements sufficient to change reluctant states' incentives. They have, however, reinforced the need for facilities and provided a legal basis for normative pressures on governments to ensure that they are provided.

* * *

Common Themes in Reporting, Enforcement, and Provision of Reception Facilities

In light of the evidence presented in the last three chapters, what general conclusions can we draw across these three disparate areas of governmental behavior? The common thread borne out by this investigation of governments' responses to efforts to improve their reporting, their enforcement, and their provision of reception facilities is that the major source of success in changing governmental behavior vis-à-vis oil pollution regulations has been ensuring that the rules are framed so that actors have the appropriate incentives, the practical ability, and the legal authority to change their behavior so that it is in compliance with treaty rules. Whether behavior is in response to the treaty's primary rule system, as is the case with reception facility provision, its compliance information system, as is the case with reporting, or its noncompliance response system, as is the case with enforcement, it has changed only when at least some actors have all three—the incentives, the ability, and the authority to change their behavior.

The major successes in altering governmental behavior have involved rules that took states' objectives as given and facilitated (or removed obstacles that inhibited) states' actions to achieve them. The MOU enforcement reporting system and the IMO reception facility surveys succeeded because they established clear reporting requirements and systems of data compilation and dissemination that provided reporters with positive benefits. Country reports to IMO on enforcement have generally only been duplicated and left undiscussed. Certainly there has been no "calling on the carpet." IMO conducts little analysis of the information on reporting, enforcement and reception facilities that it does receive, and therefore provides little basis on which governments or nongovernmental organizations that might have incentives to sanction or shame noncompliant governments might do so.

Enforcement of oil pollution regulations has been improved not through previously reluctant oil-loading states' undertaking the inspections needed to successfully monitor compliance with the LOT rule and the 1/15,000th provision of the 1969 amendments to OILPOL, nor through dramatic increases in developed states' aerial surveillance programs due to international requirements, but rather through establishing equipment standards whose violation could be readily detected in any port and simultaneously providing port states with the right to inspect and detain ships found to be in violation. The fact that few countries have, in the event of violation, detained ships makes it clear that states lacking interests and incentives to enforce MARPOL have not had those interests or incentives altered by the obligation to detain ships posing risks to the marine environment. MARPOL's compliance system achieved stiffer enforcement by removing the legal obstacles that prevented states like the United States from exercising their interests in enforcement. The MOU improved enforcement by creating a real-time data system that, once up and running, provided a positive benefit to its users by making enforcement easier for those already inclined and able to conduct inspections.

Similarly, provision of reception facilities has probably been increased most by IMO's approval and promotion of less costly "floating reception facilities" and growing domestic concern that removed the large costs that deterred many states from installing them prior to the 1980s. Certainly IMO did not provide funding for states to install reception facilities,[52] nor did it produce "blacklists" of countries failing to provide reception facilities. French calls to waive discharge standards for tankers going to ports without reception facilities were rejected. The states installing reception facilities have tended to be those that one would independently predict would have strong domestic environmental interests and the resources to dedicate to such installations. Those lacking strong interests, including most oil-loading states, have continued to disregard the reception facility requirements despite new, low-cost means of complying and large revenues available from the oil industry itself.

In the analysis here I have sought to identify whether rule changes can alter states' behavior. By defining the research problem in this way, I have

---

52. This contrasts sharply with the large amount of energy spent discussing the Montreal Protocol funding mechanism and with the discussions of funding mechanisms in the climate change negotiations (*Vienna Convention for the Protection of the Ozone Layer*, 22 March 1985, T.I.A.S. no. 11097, reprinted in 26 I.L.M. 1529 (1987); *Montreal Protocol on Substances That Deplete the Ozone Layer*, 16 September 1987, reprinted in 26 I.L.M. 1541 (1987); *London Amendments to the Montreal Protocol on Substances that Deplete the Ozone Layer*, 29 June 1990, reprinted in 30 I.L.M. 539 (1991); and *Framework Convention on Climate Change*, UN Document A/AC.237/18 (Part II)/Add. 1 (15 May 1992).

essentially precluded a finding that rules alter behavior by changing the goals and objectives of states; this pathway would predictably have so gradual an effect that it would prove difficult to clearly attribute any of that effect to the rule change. Rules and processes that did not directly alter behavior might have nonetheless changed behavior through less direct pathways, such as through changing states' perceptions of goals. Research designed to isolate clear links between a single independent variable and a dependent variable may miss causal connections that are incapable of being isolated. Our inability to isolate rules as a strong influence on behavior may mean that no such influence exists. It may also mean that the influence exists, but cannot be isolated. There are more reception facilities today than in 1954. More states conduct aerial surveillance and port inspection programs than in 1954. The increases in these activities cannot be traced directly to new international rules; yet, at an impressionistic level, it seems likely that new rules on reporting, enforcement, and the provision of reception facilities have been not only reflected by, but have contributed to an increase in concern and an alteration of interests that have, in turn, led to new behaviors. Although avoiding exclusive reliance on such impressions is precisely the major empirical goal of this study, they may nonetheless be suggestive of causal links that cannot be isolated and proven in a muddy, multicausal world.

Many analysts have argued that international treaties most influence states' behavior through a process of information sharing and application of normative pressures that slowly and subtly, but nonetheless forcefully, alter states' perception of their goals and interests. This premise is generally the basis underlying Haas's epistemic community argument,[53] Levy's view of the LRTAP regime,[54] and Young's view of international cooperation.[55] Some degree of goal alteration has undoubtedly occurred in the case of oil pollution. The various marine pollution conferences and

53. Haas, *Saving the Mediterranean.*

54. Marc Levy, "European Acid Rain: The Power of Tote-board Diplomacy," in Peter Haas, Robert O. Keohane, and Marc Levy, eds., *Institutions for the Earth: Sources of Effective International Environmental Protection* (Cambridge, MA: The MIT Press, 1993).

55. Oran Young, *International Cooperation: Building Regimes for Natural Resources and the Environment* (Ithaca, NY: Cornell University Press, 1989), chapter 3.

the regular meetings of IMCO/IMO bodies, including MEPC, have surely increased states' concern about and attention to oil pollution. The relatively gradual increase in the number of reception facilities since the 1950s and the slow but growing increase in the number of countries reporting were clear, even though they do not appear to have been well correlated with any given rule change. Difficulties in assessing whether treaties and negotiations lead states to change their goals and therefore change their behavior arise because the time periods involved prevent the exclusion of rival factors that changed during those same periods. As has been pointed out in numerous discussions of causation, the farther away in time a cause is from its alleged effect, the more difficult the task of establishing the causal link.[56] Thus, while the oil pollution regime may well have improved compliance gradually by altering states' perceptions of the benefits of reducing pollution, it would prove difficult to clearly distinguish that source of changed perceptions from exogenous domestic political forces that were also altering states' perceptions independently.

The oil pollution regime has increased reporting, enforcement, and reception facility provision by facilitating these behaviors by those governments that had preexisting incentives to undertake them. States lacked the will to positively induce compliance by other actors—e.g., by financing reception facilities—or to evaluate compliance so that violators could be shamed, if not sanctioned, into compliance. Yet MARPOL provided new international rights that allowed for more vigorous enforcement by states with incentives to enforce. MARPOL and MEPC promoted new means of complying with requirements regarding reception facilities that increased their provision by states that were so inclined. The MOU developed reporting mechanisms that have resulted in far better response rates than were seen under IMO reporting. All of these facts demonstrate that treaty provisions and corollary procedures do influence governmental behavior and can induce governments to change their behavior. Although success may be slight and slow in such efforts, it is nonetheless real.

56. See, for example, Paul W. Holland, "Statistics and Causal Inference," *Journal of the American Statistical Association* 81 (December 1986).

III

Changing Industry Behavior

Although altering government behavior has constituted an important role of the regulations established in the International Convention for the Prevention of Pollution of the Sea by Oil (OILPOL)[1] and the International Convention for the Prevention of Pollution from Ships (MARPOL),[2] these agreements have ultimately been intended to change the behavior of private tanker operators.

The initial 1954 OILPOL agreement limited the oil content (parts of oil per million parts of water) of a tanker's discharges near shore. Nations have since negotiated three major changes to the primary rules originally delineated in 1954 in the hope of improving compliance. The first changed the discharge standard with which tanker operators were required to comply from one that operators lacked the ability to monitor to one that operators could monitor and governments could verify. The second change involved adopting new limits on total discharges that analysts argued would dramatically improve enforcement.[3] The third rule

1. *International Convention for the Prevention of Pollution of the Sea by Oil,* 12 May 1954, 12 U.S.T. 2989, T.I.A.S. no. 4900, 327 U.N.T.S. 3, reprinted in 1 I.P.E. 332, hereinafter cited as *OILPOL 54.*

2. *International Convention for the Prevention of Pollution from Ships,* 2 November 1973, reprinted in 12 I.L.M. 1319 (1973), 2 I.P.E. 552; and *Protocol of 1978 Relating to the International Convention for the Prevention of Pollution from Ships,* 17 February 1978, reprinted in 17 I.L.M. 1546 (1978), 19 I.P.E. 9451, hereinafter together cited as *MARPOL 73/78.*

3. *1969 Amendments to the International Convention for the Prevention of Pollution of the Sea by Oil,* 21 October 1969, reprinted in 1 I.P.E. 366, hereinafter cited as *OILPOL 54/69.*

change required tanker owners to build and equip tankers with technologies that reduce the amount of oil-water mixtures generated. Although compliance with the equipment standards is very expensive and compliance with both the rate and total discharge standards has proven relatively cheap, the equipment standards have proven far more effective in inducing compliance than have any of the various discharge standards. As has already been illustrated in chapter 5, the discharge and equipment rules provided the foundation for two distinct regulatory strategies. As separate provisions of the same international treaty, they applied to the same tankers and the same countries. However, they depended on quite different compliance systems to prevent tankers from discharging at sea. This section describes both strategies, detailing the incentives for compliance, the expected influences of each compliance system, and the impact each compliance system had on industry behavior. Chapter 7 evaluates the discharge approach, while chapter 8 focuses on the equipment approach.

7
Discharge Limits: A Failed Strategy

The thirty-two nations that negotiated the 1954 OILPOL agreement sought to reduce ocean oil pollution by limiting how much oil ship operators could discharge overboard. Despite differences between the content, rate, and total discharge limits, all three have relied on a compliance system founded on a deterrent model of compliance: to deter ship operators from making illegal discharges, governments had to create a system that made it very likely that such discharges would be successfully detected and their perpetrators prosecuted, with high costs imposed on those that were convicted. The system relied on increasing the likelihood and magnitude of sanctions to decrease the incentives to illegally discharge oil in the first place.

The 1954 agreement was a compromise between the United Kingdom and a few allies, which advocated stringent regulation in response to strong domestic environmental groups, and most other states, which viewed any regulations as environmentally unnecessary and detrimental to their industries.[4] These latter states' opposition to the British objective of ocean-wide limits resulted in a final compromise that prohibited discharges of over 100 parts per million (ppm) made within fifty miles of shore.[5] The British subsequently negotiated the 1962 amendments to OILPOL, which made the 100 ppm standard apply ocean-wide for all new tankers.

By the late 1960s the perception that the 1954 convention, and especially its 1962 amendments, would not induce compliance produced

4. Larger zones were established in the North Sea, off northwestern Europe, and off Australia; see Article III(1), *OILPOL 54*.

5. See chapter 3 for a more extended discussion of the politics of this period.

pressures to revise the regulations fundamentally. These pressures came from two different quarters that held quite different views of the source of noncompliance and the appropriate remedy. On the one hand, oil and shipping interests, supported by the now less activist United Kingdom, as well as France, the Netherlands, and Norway, contended that non-compliance was likely because tankers lacked the practical ability to comply. The 100 ppm standard did not correspond to existing technical capacities to separate oil from water or to monitor the resulting dis-charges.[6] Reliable oil content meters to monitor compliance with the 100 ppm standard would not become available until the late 1980s.[7] Oil company representatives also argued that adequate equipment, even if available, "would be prohibitively expensive" and was unnecessary if a "more realistic and sophisticated" rate standard was adopted.[8] A stan-dard measured in liters per mile rather than parts per million had the advantage that it could be cheaply monitored using a tanker's existing equipment, thereby legitimizing the use of the load on top (LOT) proce-dures that oil companies had been promoting while eliminating the need to develop and install expensive new equipment.[9] While industry repre-sentatives assured governments that LOT would reduce pollution, they

6. One expert "found no evidence to inform us how the original value of 100 ppm was chosen"; see J. Wardley-Smith quoted in K. G. Brummage, "The Problem of Sea Pollution by Oil," in Peter Hepple, ed., *Pollution Prevention* (London, England: Institute of Petroleum, 1968), 99.

7. E. J. M. Ball, the head of the Oil Companies International Marine Forum, noted that even in the 1990s "to the conscientious operator, the lack of a reliable oil content meter remains a problem" (Interview, London, England, 26 June 1991); and E. Somers, "The Role of the Courts in the Enforcement of Environ-mental Rules," *International Journal of Estuarine and Coastal Law* 5 (February 1990), 196. See also Brummage, "The Problem of Sea Pollution," 99; and William T. Burke, Richard Legatski, and William W. Woodhead, *National and International Law Enforcement in the Ocean* (Seattle: University of Washington Press, 1975), 200.

8. W. M. Kluss, "Prevention of Sea Pollution in Normal Tanker Operations," in Hepple, *Pollution Prevention*, 114.

9. See J. H. Kirby, "The Clean Seas Code: A Practical Cure of Operational Pollution," *International Conference on Oil Pollution of the Sea* (Rome: October 1968); and J. H. Kirby, "Background to Progress: Interview with Sir Gilmour Jenkins," *The Shell Magazine* 45 (January 1965).

opposed compulsory compliance, instead seeking "*voluntary* adoption of regulations [as] the eventual goal."[10]

On the other hand, a growing scientific consensus that crude oil discharged beyond fifty miles from shore could still soil beaches, as well as growing environmentalism, especially in the United States, and the 1967 grounding of the *Torrey Canyon* were creating strong pressures to develop more stringent regulations.[11] Those concerned with marine environmental protection took the skeptical view that noncompliance was the result of an absence of incentives to comply, which was due, in turn, to poor enforcement. Evidence available on enforcement and compliance, from a 1962 Intergovernmental Maritime Consultative Organization (IMCO) survey and increasing pollution of beaches, suggested to many that tankers would simply not retain oil slops on board unless enforcement of rules regulating discharges improved.[12] The United States and other members of IMCO's Subcommittee on Oil Pollution (SCOP) had seen little evidence of behavior change by industry and sought stronger "methods of detection of pollution and enforcement of the [1954/62] Convention."[13] The United States, the Soviet Union, Japan, Sweden, and Germany opposed the industry preference for voluntary provisions, pushing instead for specific mandatory limits on total discharges.[14]

10. C. T. Sutton, "The Problem of Preventing Pollution of the Sea by Oil," *BP Magazine* 14 (Winter 1964), 10. Emphasis added. None of the early industry arguments for LOT mention enforcement or inspection as a benefit of the system. See, for example, OP I/21 (15 January 1965); Kirby, "Background to Progress"; Kluss, "Prevention of Sea Pollution"; and Brummage, "The Problem of Sea Pollution."

11. R. Michael M'Gonigle and Mark W. Zacher, *Pollution, Politics, and International Law: Tankers at Sea* (Berkeley: University of California Press, 1979), 100; and Sonia Zaide Pritchard, *Oil Pollution Control* (London: Croom Helm, 1987), 130–131.

12. Pritchard, *Oil Pollution Control*, 112; and M'Gonigle and Zacher, *Pollution, Politics, and International Law*, 220–221.

13. D. Young quoted in Brummage, "The Problem of Sea Pollution," 97; and G. Victory, "Avoidance of Accidental and Deliberate Pollution," *Coastal Water Pollution: Pollution of the Sea by Oil Spills* (Brussels, Belgium: Committee on the Challenges of Modern Society of NATO, 1970), 2.2.

14. Japan disliked LOT because combining slops with crude oil increased the latter's salt content—because of the slops' prolonged contact with saltwater—which made refining more difficult and costly. The Soviet Union opposed LOT

These competing views led, in 1969, to a unanimously approved compromise. The preferences for more stringent mandatory standards were defined in terms that permitted tankers to comply by using the industry-preferred method of LOT: near shore, discharges of "clean ballast" had to leave no visible trace; elsewhere discharges could not exceed sixty liters per mile (60 l/m); and total discharges per ballast voyage could not exceed 1/15,000th of a tanker's capacity.[15] These standards promised to improve compliance by improving the ability of tankers to self-monitor discharges and the ability of governments to verify compliance. Supporters of LOT accepted the mandatory and stringent requirements as a means to legitimize the use of LOT while averting equipment requirements. Others saw the new regulations as more enforceable and more environmental. The "clean ballast" rule meant that "any sighting of a discharge from a tanker . . . would be much more likely to be evidence of a contravention."[16] The total discharge rule meant that port authorities could assume any tanker with clean tanks had blatantly violated the agreement.[17] More important, the total ballast provision was the first international requirement for tankers to reduce their discharges rather than merely redistributing them. The amendments did not merely codify industry preferences or represent a lowest common denominator among the bargaining nations, but were a clear middle ground between the newly environmentalist United States and the countries representing oil industry interests, led by the United Kingdom. These limits were subsequently incorporated into the 1973 MARPOL, where they were also strengthened to 1/30,000th of the total cargo-carrying capacity (tcc) for new tankers.[18]

---

because most of its tankers worked short voyages that provided insufficient time for oil-water separation, so slops had to be retained on board. There are few details of exact positions on the 1/15,000th rule, as opposed to positions with respect to LOT itself, however. This account relies extensively on M'Gonigle and Zacher, *Pollution, Politics, and International Law,* 101–102.

15. Article III, *OILPOL 54/69.*

16. Resolution A.391(X) (1 December 1977), Annex, par. 5.

17. Kirby, "The Clean Seas Code," 200 and 209; and Burke, Legatski, and Woodhead, *National and International Law Enforcement,* 129.

18. The language was altered slightly in MARPOL to "1/15,000th of the total quantity of the particular cargo of which the residues formed a part" (Annex I,

First-Order Compliance Incentives

The negotiators settled on these standards precisely because "any responsibly run ship, no matter how big, could operate" within them if it used LOT.[19] Tankers could ensure they complied if they wanted to. Indeed, a major oil industry argument for LOT was the claim that 80 percent of the world tanker fleet was already using it to reduce oil discharges.[20] Tests by Mobil, Exxon, Norwegian shipping interests, and the Japanese government all showed that LOT could keep discharges on both small and large tankers at least below the 1/15,000th level, and in some cases below the 1/50,000th level.[21] Beyond the fifty-mile zones, captains could readily calculate discharge rates from existing equipment that measured ship speed and discharge flow. And tanker personnel could avoid making illegal discharges within the prohibition zones simply by halting the discharge of water that had been separated from slops as soon as the first traces of oil appeared on the ocean surface.

Having the ability to comply did not mean all tanker operators had incentives to do so, however. A tanker captain's "first-order" incentives to reduce oil discharges, i.e., incentives independent of deterrent threats, depended on the costs of recovering waste oil, the value of that oil, and the ownership of the oil being transported. The incentives to use LOT effectively enough to meet the 1969 standards differed between independents and oil company tankers. Independents carrying oil on charter got paid for the amount of oil on-loaded, i.e., the bill of lading weight, not the amount delivered. Since the oil had already been paid for, discharging oil at sea on the ballast voyage cost an independent nothing. Indeed, using LOT reduced the cargo room and the bill of lading weight

---

Regulation 9(1)(a)(v), *MARPOL 73/78*). New tankers were defined as those for which building contracts had been drawn up after December 1975, whose keels had been laid after June 1976, or which had been delivered after December 1979 (Annex I, Regulation 1(6), *MARPOL 73/78*).

19. Kirby, "The Clean Seas Code," 208.

20. Kirby, "The Clean Seas Code."

21. OP I/21 (15 January 1965), 3; and William Gray, "Testimony," in U.S. House of Representatives, Committee on Government Operations, *Oil Tanker Pollution—Hearings: July 18 and 19, 1978* (95th Congress, 2nd session) (Washington, DC: GPO, 1978), 10–11.

by the amount of slops retained, thereby actually reducing the payment an independent received.[22] On short or stormy voyages, the inability to separate and discharge most of the water in the slops reduced cargo room still further. The market itself produced direct disincentives to independents' using LOT. Some refineries refused to accept cargoes that had been combined with slops because such cargoes had a higher salt content, which made refining more difficult. Some countries required tankers to arrive with no slops on board to ensure that no discharges polluted their ports. Only if the cargo owner reimbursed the independent would the independent have any incentives to adopt LOT.

In contrast, for oil companies and other cargo owners the benefit of having all the oil they paid for actually delivered offset the slightly smaller cargo capacity of an LOT tanker. At 1976 prices, the lower bill of lading weight cost the tanker owner some $700, while the value of oil recovered benefited the cargo owner some $16,000.[23] Indeed, these benefits were sufficient that British Petroleum and Shell reimbursed independent operators for the direct costs of LOT for a period of time.[24] Although these companies could gain from cooperating to pay for compliance by independents and thereby prevent the adoption of equipment rules, each had incentives to let others pay the costs, and the system of reimbursement was soon discontinued.[25]

In general, however, independent tanker owners had few incentives to use LOT, while oil companies, which own both tankers and the oil they carry, have strong incentives to use LOT. Although other methods were available to help tankers conserve oil, LOT required no new equipment and was by far the most cost-effective retention option.[26] It followed that,

22. This discussion of LOT costs draws extensively on William G. Waters, Trevor D. Heaver, and T. Verrier, *Oil Pollution from Tanker Operations: Causes, Costs, Controls* (Vancouver: Center for Transportation Studies, 1980), 85–89.

23. Waters, Heaver, and Verrier, *Oil Pollution,* 124.

24. Kirby, "Background to Progress," 26–27.

25. MEPC 32/10 (15 August 1991). On the theoretical point, see Arthur A. Stein, "Coordination and Collaboration: Regimes in an Anarchic World," in Stephen D. Krasner, ed., *International Regimes* (Ithaca, NY: Cornell University Press, 1983).

26. See, for example, Waters, Heaver, and Verrier, *Oil Pollution*; Charles S. Pearson, *International Marine Environmental Policy: The Economic Dimension*

as oil prices rose or new technologies became available or less expensive, these incentives to reduce discharges should have grown and compliance should have become more likely.[27] The oil price increases of 1973 and 1978 greatly increased the value of the conserved oil and the incentives of cargo owners to use LOT. At any given time, however, oil transporters that viewed recovery as uneconomical would have incentives to violate international restrictions on discharges. The costs and benefits of retaining slops on board provided the major first-order incentives to comply with the total discharge standards that took effect in 1978.

Large oil companies had certain additional incentives to adopt pollution reduction technologies independent of international action. As public concern increased over oil pollution in the wake of major spills like that of the *Torrey Canyon*, oil companies increasingly believed that their reputations and long-term profits were contingent on being "good citizens" and adopting pollution control technologies. Even absent any direct enforcement activity by governments, such market pressures could lead large oil companies to adopt more environmentally sound practices. Even to the extent that these forces influenced large corporate actors, they would have little influence on the many small oil transport companies that had little recognition among the general public and were less susceptible to such public pressures.

While oil companies had incentives to comply with the limits on total discharges, neither they nor independents had economic incentives to comply with the original 1954 oil content limits or the 1969 clean ballast and discharge rate standards. The new standards increased the ability of operators to monitor their discharges, but did not change their incentives to do so. Compliance with either standard still required tankers to make all discharges containing oil outside the prohibition zones. Since a tanker's route generally included considerable stretches of ocean where discharges were less restricted, as in the middle of the Mediterranean,

---

(Baltimore: The Johns Hopkins University Press, 1975), chapter 4; Philip A. Cummins, Dennis E. Logue, Robert D. Tollison, and Thomas D. Willett, "Oil Tanker Pollution Control: Design Criteria vs. Effective Liability Assessment," *Journal of Maritime Law and Commerce* 7 (October 1975); MEPC V/Inf. 4 (15 March 1976); MEPC VI/Inf. 7 (6 September 1976); and MEPC VIII/Inf. 16 (5 December 1977).

27. See figure 3.2 for crude oil prices between 1945 and 1990.

complying with the zonal prohibitions was quite cheap. In contrast, applying the 60 l/m standards throughout the oceans and limiting total discharges significantly increased a tanker's costs of complying. Whereas before such tankers could have discharged their wastes while en route to their next port, tankers that wanted to comply now faced long delays in ports to discharge slops or had to retain slops on board until they arrived in a port with reception facilities. Therefore, even tankers that had complied with the 100 ppm standard and intending to comply with the clean ballast rule would have few incentives to meet the more stringent 60 l/m and total discharge standards.

In general, then, no tankers had any significant independent incentives to comply with the content and rate standards, and independent tankers lacked incentives to comply with the total discharge standards; compliance therefore depended on effective deterrence through identifying and responding to violations. Given the obvious private incentives to violate the discharge standards, the success of the subregime's compliance system depended on creating strong second-order incentives to comply in the form of credible and potent deterrent threats.

Expected Pathways of Influence

The new rules were expected to improve the compliance system in three ways: making compliance easier and cheaper, making detection of violations easier, and making identification of violators and collecting evidence on them more reliable. The new rules improved the primary rule system by making it cheaper and easier for a tanker captain interested in complying to monitor his own compliance. Without installing any new equipment, a tanker captain could use LOT to separate the oil and water in the ballast and tank-cleaning slops, discharge the water from underneath the oil, stop the discharge as soon as oil appeared in the discharge stream, and consolidate all remaining oil for delivery as part of the next load of cargo. A tanker captain could also readily track the total amount of oil discharged, although the 1/15,000th limit was so low as effectively to prohibit any discharges of oil and require that all slops be retained on board. None of these changes made the primary rules more specific; the "no visible trace," 60 l/m, and 1/15,000th rules did not distinguish compliance from noncompliance any more precisely than the 100 ppm standard. They did, however, accord better with tankers' existing capacities to self-monitor behavior.

More important, the 1969 amendments had "that essential virtue of any worthwhile legislation—the possibility of enforcement."[28] All three components promised to make the compliance information system more effective in detecting violations. The clean ballast rule addressed the specific detection problems of the 100 ppm rule. As James Kirby, LOT's major proponent, noted in 1968, "it is quite possible to cause a visible film behind a ship even though the oil content of the effluent is well below 100 ppm. It is this fact basically which renders the present [1954/62] Convention quite impossible to police since it cannot be proved by observation of a streaming film that the ship was exceeding the present [100 ppm] definition of pollution."[29] In contrast, under the clean ballast rule a visible oil slick was, by definition, a violation.[30]

Verifying violations of the 60 l/m standard proved more difficult. This could not be measured independent of the equipment on board the violating tanker. The 1954 convention had required that tankers log all oil discharges, including illegal ones, in an oil record book (ORB) to be kept on board. The version of the ORB incorporated in the 1962 amendments to OILPOL required only that tanker personnel record the position and "approximate quantities" of oil wastes discharged. Since no devices to measure compliance with the 100 ppm standard existed, the initial ORB did not request oil content information. With the adoption of the 60 l/m standard, tankers could be assumed to have adequate measurement equipment on board. Negotiators therefore also revised the ORB to include the tanker's speed, the time and the ship's location at the

28. M. P. Holdsworth, "Convention on Oil Pollution Amended," *Marine Pollution Bulletin* 1 (November 1970), 168.

29. Kirby, "The Clean Seas Code," 209.

30. Ton IJlstra, "Enforcement of MARPOL: Deficient or Impossible," *Marine Pollution Bulletin* 20 (December 1989). In 1993 the MEPC passed a resolution revising the guidelines for visibility limits on oil discharges in an attempt to improve the legal nature of evidence collected from aerial surveillance planes using high-technology radar systems. Whether these will increase the likelihood of successful prosecution remains to be seen. See MEPC 34/23 (6 August 1993); MEPC 33/4/6 (31 July 1992); and MEPC 33/Inf. 28 (31 July 1992).

beginning and end of each discharge, and the rate of flow of the discharge. A captain now had to record the information that would allow port authorities to independently verify a violation after the fact. While still based on the dubious assumption that tanker operators would willingly incriminate themselves, the new oil record book at least was an improvement on one that required information having little relation to the established discharge standards.

Given the reliance on self-incrimination, it is surprising to see the importance policymakers have placed on the oil record book as a means of detecting violations. In 1962 the United States Secretary of State stressed that the oil record book was a "vital part of the Convention."[31] While oil company representatives noted that captains could easily enter false data in the ORB without fear of detection, they still claimed that "the Oil Record Book is taken very seriously."[32] This faith in the ORB resulted in regular revisions in 1962, 1969, 1973, and 1978. As recently as 1987, the U.S. Coast Guard recommended requiring a similar log book to reinforce MARPOL's provisions on garbage discharged from ships.[33]

Although one would imagine that tanker captains who were knowingly conducting illegal discharges would not log such discharges in their oil record books, oil record books have served as the basis for detecting violations.[34] Usually they have been used to collect supporting evidence

31. Anonymous, "Tough Measures against Oil," *Marine Pollution Bulletin* 1 (May 1970).

32. LeCoq of Esso Petroleum quoted in Brummage, "The Problem of Sea Pollution," 99. See also Kluss, "Prevention of Sea Pollution," 119; Pritchard, *Oil Pollution Control,* 111; Cummins et al., "Oil Tanker Pollution Control," 177; and Burke, Legatski, and Woodhead, *National and International Law Enforcement,* 201.

33. Rear Admiral John William Kime, "Statement," in United States House of Representatives, Committee on Merchant Marine and Fisheries, *Report on Hearings on Annex V: Regulations for the Prevention of Pollution by Garbage from Ships to the 1978 Protocol Relating to the International Convention for the Prevention of Pollution from Ships,* Executive Report no. 100–108 (96th Congress; 2nd session) (Washington, DC: GPO, 14 October 1987).

34. See, for example, German enforcement reports to IMO contained in MEPC 30/17 (20 July 1990) and MEPC 32/14 (13 November 1991).

to confirm violations detected by other means. A ship discovered while discharging would be tracked to port and its ORB inspected to provide additional evidence that the discharge had occurred at the time observed. The obvious incentives against self-incrimination have made naval and aerial surveillance programs the more common means of detecting violations of the clean ballast and 60 l/m criteria. These programs must patrol wide areas of ocean to detect slicks that are often difficult to link with particular tankers. It has been observed that "the difficulties involved in patrolling the ocean are a function of manpower versus space. Therefore, the chance of catching a ship in the act of an illegal discharge is minimal."[35] Initiating legal action usually relied on having a sufficiently large slick to serve as *prima facie* evidence of an illegal discharge.

Whether a government instituted a surveillance program depended both on abilities and incentives. Many countries simply lacked the ships, aircraft, and operating funds needed to conduct a surveillance program that had any significant likelihood of detecting (and hence deterring) violations. Even most states with sufficient resources have not created ongoing and extensive monitoring programs, lacking either concern regarding oil pollution or faith that dedicating resources to such a program would actually result in significant deterrence. The huge effort needed to monitor even a small fraction of the potential areas in which and times during which violations could be committed made such programs an unattractive use of pollution control funds.

As was noted in chapter 5, the 1960s saw most countries completely failing to conduct any enforcement efforts at all. In the 1970s many developed countries established aerial surveillance programs, but most of these have remained relatively small in size and have since decreased in size or been discontinued. In reports to IMO in the 1980s, only seventeen of over sixty-five parties to MARPOL reported having any programs for detecting discharges at sea.[36] Port inspections of oil record books appear

35. Jeff B. Curtis, "Vessel-source Oil Pollution and MARPOL 73/78: An International Success Story?" *Environmental Law* 15 (1985), 707.

36. Derived from Gerard Peet, *Operational Discharges from Ships: An Evaluation of the Application of the Discharge Provisions of the MARPOL Conven-*

to have been even less frequent, with less than ten countries reporting discrepancies in these books to IMO during this period.[37]

Experts were particularly optimistic regarding improved enforcement of the total discharge standards. As early as 1954, the benefits of total prohibitions relative to zones and discharge flow limits had been pointed out: "It would be much easier to detect violations of a total prohibition, where the absence of a proper explanation as to how oil residues had been disposed of would provide evidence of an offense."[38] Total discharge limits substituted in-port inspections for surveillance programs, eliminating both the costs of patrolling large areas and the problem of "passive-voice violations," violations which had clearly occurred, but for which no ship could be identified as the culprit. Although the new ORB required entry of the total cargo-carrying capacity of the ship and the amounts of oil discharged during each operation, the total discharge standard eliminated exclusive reliance on a system based on such self-incrimination.[39] As opposed to discharge flow limits, violations need not be detected while in process and could be verified in any oil-loading port.

Kirby argued that LOT would be "automatically enforced worldwide" and that a tanker would have to choose between complying or going out of business.[40] The Oil Companies' International Marine Forum (OCIMF) proposed that oil company personnel extend existing procedures by inspecting a tanker before it loaded new cargo and after its ballast voyage. It was said that "direct measurement of the 1/15,000th standard is almost impossible"[41] because the slops a tanker should have on board depended on the degree of evaporation and clingage and on

---

tion by Its Contracting Parties (Amsterdam: AIDEnvironment, 1992), Annex 5 and Annex 10.

37. Data compiled by author from Marine Environment Protection Committee (MEPC) enforcement documents.

38. Lord Runciman quoted in Pritchard, *Oil Pollution Control,* 102.

39. See, for example, Victory, "Avoidance of Accidental and Deliberate Pollution," 2.4; and G. Boos, "Critical View of 1969 Amendments," *Marine Pollution Bulletin* 1 (November 1970), 170.

40. Kirby, "The Clean Seas Code," 212.

41. Interview, Nigel Scully, British delegate to IMO, London, England, 5 July 1991.

what technologies the tanker had installed to reduce slops, such as crude oil washing and coated tanks.[42] Nonetheless, identifying tankers that arrived with clean tanks and no slops "would reveal at least the more blatant infractions against discharge rules."[43] OCIMF and the International Chamber of Shipping (ICS) later published a pamphlet on how oil company inspectors could provide information to the port state for referral to the flag state for investigation and prosecution. For oil companies, this proposal had the virtue of not increasing government regulation while providing a means for them to pressure independents into using LOT to preserve precious cargo. Although several developed countries supported the idea, no OPEC countries offered to participate.[44]

Even after MARPOL expanded the right to conduct in-port inspections from flag states and oil companies to port states, inspections for total discharge violations have remained extremely rare. The oil-exporting states that had the capacity to inspect showed that they had no incentives to inspect. Most OPEC states are not parties to OILPOL and MARPOL, and those that are appear to have no incentives to undertake such inspections. Oil companies that did conduct such inspections during the 1970s were neither prompted nor felt obligated to reveal tanker-by-tanker information on total discharges. Industry representatives quickly pointed out that the large discharges they identified were not violations, because the total discharge standard would not take effect in 1978, the surveys having been discontinued the previous year.[45]

The amendments adopted in 1969 also promised to improve OILPOL's noncompliance response system by making it easier to produce legally

42. Martin Holdsworth cited in Sonia Zaide Pritchard, "Load on Top: From the Sublime to the Absurd," *Journal of Maritime Law and Commerce* 9 (January 1978), 215; and Gray, "Testimony," 92.

43. Burke, Legatski, and Woodhead, *National and International Law Enforcement,* 129; and Kirby, "The Clean Seas Code," 211.

44. Such proposals were discussed extensively at the first, second, and third meetings of the MEPC. See, for example, MEPC I/4/1 (20 February 1974); MEPC II/WP.2 (19 November 1974); MEPC II/4 (2 October 1974); MEPC II/4/1 (14 November 1974); MEPC III/10 (16 May 1975); and MEPC III/10/Add. 1 (5 June 1975).

45. Gray, "Testimony."

sufficient evidence of violations and to link violations to violators. Violation of the clean ballast provision could, at least plausibly, be proved by photographic evidence provided in court. Violation of the 60 l/m rate standard was less independently verifiable, but could be demonstrated in court if the captain and crew had kept an accurate oil record book as required by MARPOL. The total discharge standard held the most promise, however, since clean tanks would be "prima facie evidence" of a violation.[46] The onus would be on the tanker to prove that "she had carried out conscientiously a proper oil conservation procedure while at sea."[47] Additionally, as was not the case with the two discharge flow regulations, detecting a violation always involved positive identification of the violator.

These "improvements" have not actually enhanced enforcement for a variety of reasons. All three rules still relied on the flag state to prosecute and convict those responsible for tankers caught violating discharge rules beyond territorial waters. Only if a tanker discharged illegally within a state's twelve-mile territorial sea and then entered a port of that state could that state prosecute those responsible for a tanker flagged elsewhere. As was delineated more fully in chapter 5, flag states have generally not aggressively followed up on evidence referred to them.[48] Flag-of-convenience states often lack both the ability and the incentives to prosecute. They lack the ability since tankers flying their flags may rarely enter their ports. They lack incentives because vigorous unilateral enforcement will prompt tankers to move their flags, and the large associated registry fees, to less scrupulous states.[49]

In addition, even states sincerely seeking to prosecute violations of the discharge process standards—100 ppm, clean ballast, and 60 l/m—face

46. See Kirby, "The Clean Seas Code," 200; and Pritchard, "Load on Top," 213.

47. Kirby, "The Clean Seas Code," 209. See also Victory, "Avoidance of Accidental and Deliberate Pollution," 2.3.

48. See, for example, Organization for Economic Cooperation and Development, "OECD Study on Flags of Convenience," *Journal of Maritime Law and Commerce* 4 (January 1973).

49. See Paul Stephen Dempsey, "Compliance and Enforcement in International Law: Oil Pollution of the Marine Environment by Ocean Vessels," *Northwestern Journal of International Law and Business* 6 (1984), 526. Registry revenues can make up a significant share of government revenues for some of the poorest flag states.

obstacles. Evidence of a violation often does not point to a violator, and otherwise convincing evidence often fails to meet the legal standards of proof needed to prosecute and convict. Aggressive states also find it difficult to impose penalties on convicted violators sufficient to deter them from committing future violations.

The fact that there are no private economic incentives for independent tankers to comply with any discharge standards and for oil company tankers to comply with the content and rate standards makes effective deterrence essential for compliance. Yet the evidence just delineated strongly suggest that the information and response systems do not create the credible and potent threat necessary for deterrence.

Drawing on the evidence laid out in chapter 5, we can develop a sense of how likely a tanker operator would be to be deterred by the compliance systems associated with the discharge provisions. The use of aerial surveillance with sophisticated radar can make detecting an oil slick reasonably likely. But the mismatch between resources applied to surveillance and the potential times and locations of violations causes even many daytime discharges to go undetected. For example, the Dutch agency responsible for aerial surveillance estimates that it detects only half of the slicks that are actually discharged. Many states do not have such programs, so their rates of effective detection are much lower, especially on the high seas. Once such a slick is detected, there is still a strong possibility that the violator cannot be identified. British and Dutch data suggest that only some 10 to 30 percent of detected slicks can be linked to ships.[50] A captain intent on discharging illegally can discharge at night or in stormy weather and almost assuredly avoid detection. Of all the cases referred to flag states, the likelihood that the responsible tanker would be prosecuted and convicted has remained consistently below 20 percent since the 1970s. Even if we use the upper limits of each of these estimates, it is clear that on average only some 3 percent of all those tankers responsible for making illegal discharges are successfully prosecuted.[51] Even if the probabilities of both identifying responsible

50. See chapter 5.

51. This 3% assumes a 50% probability that the slick is detected, a 30% probability that it can be linked with a ship, and a 20% probability that the case can be successfully prosecuted.

tankers and successfully prosecuting those responsible for them were increased to 50 percent, those responsible for a violating tanker would have only a 13 percent chance of being convicted.

These low probabilities might nonetheless prove adequate to deter tanker captains from making such discharges if nations imposed fines that were high enough that the expected costs of discharging at sea exceeded the benefits. All available evidence demonstrates, however, that most states impose fines that are "very low in comparison to the price the vessel would have to pay for using port reception facilities."[52] The various efforts made under OILPOL and MARPOL to induce nations to increase their average fines for discharge violations have proved largely futile (see chapter 5). Average fines reported to IMO have never exceeded $7,000 and have usually been considerably below that level. Using the economist's concept of expected value, we can multiply the 3 percent figure calculated above and this average fine to identify the expected cost of violating any of the discharge process standards as only $200. Even taking the 13 percent figure increases this to only $900. This figure is close to the opportunity costs for a single hour of a tanker's time.[53] Yet compliance involved several hours of a tanker's time, and the attendant costs, to discharge into reception facilities. Further exacerbating the problem, "shipowners and their insurers routinely indemnify the masters of their ships against fines imposed upon them for oil pollution."[54] In short, a tanker captain evaluating the expected costs of violating OILPOL or MARPOL's discharge standards finds both the magnitude and likelihood of a penalty for violation to be quite small relative to the costs of compliance.

The total discharge standards held out the promise of eliminating many of these obstacles to effective deterrence; systematic inspections could make the detection rate close to 100 percent, detecting a violation was equivalent to identifying the violator, and evidence of violations would have made prosecution relatively easy. None of this promise was realized,

52. MEPC 29/10/3 (15 January 1990).

53. Waters, Heaver, and Verrier, *Oil Pollution,* 124.

54. Owen Lomas, "The Prosecution of Marine Oil Pollution Offences and the Practice of Insuring Against Fines," *Journal of Environmental Law* 1 (1989), 54. See also MEPC 32/14/3 (17 January 1992).

however, because the inspections needed to start the legal process were never accomplished. None of the actors with the capacity to inspect for such violations have ever demonstrated that they have had any incentives to do so. To the extent that tanker companies engage in something approximating a rational calculus of whether to violate these standards, the discharge rules and the associated compliance systems have clearly failed to create conditions that would deter tankers. Although other factors may influence many tanker operators' decisions to violate or comply with these provisions, such a "rational actor" view does show that the major instrumental means by which we could expect the discharge provisions to influence tanker operators' behavior would be unlikely to be the source of much behavioral change.

Observed Compliance by Tanker Operators

The foregoing information suggests that, while some tankers, particularly oil company tankers, had some private economic incentives to practice LOT procedures effectively, most tankers had significant incentives to violate the total discharge provision and the discharge process (100 ppm, clean ballast, and 60 l/m) provisions. Equally important, it appears that the new provisions would have been unlikely to have produced any significant increase in compliance over that observed with the original 100 ppm standard. Certainly the new rules seem unlikely to have increased compliance by more successfully deterring tanker operators. These predictions of inefficacy can bolster empirical evidence of these rules' inefficacy. However, they cannot substitute for such evidence. To confirm that the rules failed to influence tanker operators' behavior requires evidence of how compliance levels have changed over time.

If the 1969 amendments to OILPOL did have any impact, it would most likely be evident after the amendments took legal effect in 1978. If they had such an impact, we would expect it to continue through the present, since the standards were not significantly strengthened upon their subsequent incorporation into the MARPOL agreement. To identify an impact of the rule change requires that we assess the prechange noncompliance level. Unfortunately the difficulties of detecting discharge violations pose obvious obstacles to academic analysis of discharge violation rates.[55]

Despite such obstacles, combining various sources of evidence can provide insight into the extent to which the clean ballast, 60 l/m, and total discharge standards have increased compliance over that elicited by the original 100 ppm standard. Several pieces of data testify to the fact that tankers commonly violated the 100 ppm standard prior to the rule change. The very pressure for revision of the standards in 1969 arose from the perception among several states that compliance was low. The various efforts to improve enforcement in the 1969 amendments, as well as in MARPOL and its protocol, reflect a fundamental belief that then-current compliance levels were lower than they should be. Many experts believed there was certainly at least a "hard core" of tankers that were freely discharging their slops into the sea.[56]

During the late 1960s and early 1970s, reports of frequent oiling of birds remained commonplace in countries concerned about oil pollution and along major trading routes, even when accidents were not reported.[57] Complaints of oil pollution, especially in the Mediterranean, increased markedly in the 1960s.[58] The increased pollution could have been due to the sharp increase in seaborne oil trade during the 1950s and 1960s, although oil industry analysts attributed it to tankers' complying with the 1954 restrictions by retaining slops while off Europe's northwest coast and discharging them in the legal zone in the middle of the Mediterranean.

More systematic data are available in a few cases. IMO data from three to four countries for 1975 through 1977 show that forty to fifty discharge violations were being prosecuted per port state, with an additional five to ten referred to flag states for action (see table 7.1). British statistics

55. Indeed policy changes that improve verifiability could well result in an increase in the number of recorded violations even while inducing a decrease in actual violations.

56. G. Victory, "IMCO and the "Load-on-Top" System," paper VI/1, in *Report of the Symposium on Prevention of Marine Pollution from Ships Held in Acapulco, Mexico on 22–31 March 1976* (London: IMCO, 1976), 36.

57. W. R. P. Bourne, "Where the Oil Came From," *Marine Pollution Bulletin* 2 (December 1971), 181.

58. Kirby, "The Clean Seas Code," 203.

on oil spills in the major oil port of Milford Haven show that the number of detected discharge incidents in port remained relatively constant between 1961 and 1974, but decreased when normalized to tanker traffic and measured in terms of tonnage or number of ships (see table 7.2). The low (1 to 4 percent) rate of spills per ship suggests that violations were infrequent. The fact that these figures are for spills in an oil-delivery port prevents us from drawing the conclusion for violations in general: tankers generally need to discharge on the voyage back to the oil-loading port on the high seas, not in port, and the relatively stringent enforcement in Milford Haven would have further bolstered a tanker's incentives to dump its oil slops out at sea. Indeed, Dutch statistics for oil spills in the North Sea show an opposite trend: spills increased from twenty-four in 1969 to almost seven hundred in 1977 (see table 7.3). Although the data include spills from sources other than tankers, do not exclude double counting of the same spill, and like the British data fail to normalize for detection effort, they provide anecdotal evidence that discharge violations were still quite common. Although none of these pieces of evidence alone would prove convincing, together they indicate that tankers still regularly discharged their slops in violation of the 1954/62 100 ppm rule.

Available evidence documents that behavior has not changed significantly since 1978. This conclusion comes from various sources of detected discharge violations, anecdotal evidence of such violations, and more general evidence of continuing discharges. Data from several sources give no indication that the number of discharge violations has decreased since 1978. National reports to IMO do not distinguish between clean ballast, 60 l/m, and total discharge violations, but do demonstrate that the aggregate number of violations has changed little over time. Violations detected by port and coastal states between 1978 and 1980 remained at the same level as the 1975 to 1977 level of forty to fifty-five violations per country, despite a steady increase in the number of countries reporting (see table 7.1). From 1983 to 1990, more countries reported fewer violations, dropping to a range of eight to nineteen per country.[59] These data show that the number of detected violations was

59. The data rely on two secondary analyses of IMO enforcement reports: the data for 1975 through 1982 come from Dempsey, "Compliance and Enforcement"; the data for 1983 through 1990 come from Peet, *Operational Discharges*.

Table 7.1
Alleged violations of the discharge provisions of OILPOL 1954/62/69 and MARPOL 73/78

	1975	1976	1977	1978	1979	1980	1981	1982	1983	1984	1985	1986	1987	1988	1989	1990
Reports of alleged discharge violations																
Number of port and coastal states reporting	3	4	4	7	10	13	5	na	10	10	13	12	10	15	15	15
Number of reported violations	145	185	166	397	497	570	91	na	87	185	199	191	113	163	181	125
Average violations per reporting country	48	46	42	57	50	44	18	na	9	19	15	16	11	11	12	8
Reports of referrals of alleged discharge violations to flag states																
Number of countries reporting	1	1	2	4	6	8	1	1	10	10	12	12	8	12	13	13
Number of reported referrals	6	9	23	115	139	180	5	10	40	150	131	170	97	144	157	98
Average referrals per reporting country	6	9	12	29	23	23	5	10	4	15	11	14	12	12	12	8

Sources: Paul Stephen Dempsey, "Compliance and Enforcement in International Law: Oil Pollution of the Marine Environment by Ocean Vessels," *Northwestern Journal of International Law and Business* 6 (1984), tables II and VIII, 487 and 511; Gerard Peet, *Operational Discharges from Ships: An Evaluation of the Application of the Discharge Provisions of the MARPOL Convention by Its Contracting Parties* (Amsterdam: AIDEnvironment, 15 January 1992), annexes 4, 5, and 11.

Table 7.2
Milford Haven oil spill statistics

	1961	1962	1963	1964	1965	1966	1967	1968	1969	1970	1971	1972	1973	1974
Number of spills detected	45	33	28	34	83	72	50	52	58	55	49	56	48	43
Number of ships in port	1,066	1,192	1,236	1,392	1,985	2,378	2,680	2,669	3,266	3,359	3,490	3,465	3,886	4,200
Number of spills per 100 ships	4.2	2.8	2.3	2.4	4.2	3.0	1.9	1.9	1.8	1.6	1.4	1.6	1.2	1.0
Tons of oil spilled (millions)	10	12	13	18	25	29	28	30	40	41	43	46	53	59
Number of spills per million tons of oil shipped	4.5	2.9	2.1	1.9	3.3	2.5	1.8	1.7	1.4	1.3	1.1	1.5	0.9	0.7

Source: G. Dudley, "The Incidence and Treatment of Oil Pollution in Oil Ports," in Jennifer M. Baker, ed., *Marine Ecology and Oil Pollution* (New York: John Wiley and Sons, 1976), 29.

Table 7.3
Dutch North Sea oil spill statistics

	'69	'70	'71	'72	'73	'74	'75	'76	'77	'78	'79	'80	'81	'82	'83	'84	'85	'86	'87	'88
Number of spills detected																				
Aerial patrol							16	39	596	363	397	294	542	340	887	528	457	282	409	318
Irregular reports	24	96	145	105	151	128	138	148	123	125	103	82	118	126	137	121	112	96	126	111
Total	24	96	145	105	151	128	154	187	719	488	500	376	660	466	1,024	649	571	381	535	429
Number of spills linked to ships									100	127	82	47	74	77	107	82	72	62	70	
Spills linked to ships as a percentage of all spills									14	26	16	13	11	17	10	13	13	16	13	
Hours of aerial patrol							275	500	629	809	848	740	520	580	492	582	574	588	608	
Spills detected per patrol hour															2.4	1.3	1.2	0.9	1.2	
Spills linked to ships per patrol hour															0.12	0.06	0.06	0.08	0.13	

Source: N. Smit-Kroes, *Harmonisatie Noordzeebeleid: Brief van de Minister van Verkeer en Waterstaat* (Tweede Kamer der Staten-Generaal: 17–408) (The Hague: Government Printing Office of the Netherlands, 1988); and C. J. Camphuysen, *Beached Bird Surveys in the Netherlands, 1915–1988: Seabird Mortality in the Southern North Sea since the Early Days of Oil Pollution* (Amsterdam: Werkgroep Nordzee, 1989), 41, 248.

consistently 65 to 75 percent lower after 1983 than before 1981. They also show, however, that the number of reported violations has remained relatively constant at one hundred to two hundred per year since 1983. The number of violations referred to flag states shows no such decrease subsequent to the 1959 amendments' entry into force in 1978. The average number of referrals per reporting country ranged from six and twelve from 1975 to 1977, between twenty-three to twenty-nine from 1978 to 1980, and from four to fifteen from 1983 to 1990. This data set indicates that there was some improvement but it is inconclusive since the decrease in number of reported violations could also indicate a decreased monitoring and enforcement effort and/or the greater success of captains in avoiding detection.

Dutch North Sea statistics show a relatively constant fifty to one hundred spills linked to ships each year from 1977 to 1987, and this represented a relatively constant percentage of all spills. Although figures are not available from prior years for comparison, the overall number of spills detected increased dramatically beginning in 1977. Part of this increase was due to an aerial surveillance program begun in 1975, whose sightings complemented the prior sightings by government ships and private observers. Although the data are difficult to compare to the data of the period before 1978, they do suggest that spills in general, and those specifically linked to ships, have remained commonplace off the Dutch coast even after the 1969 amendments took effect. British data from 1978 to 1980 confirm this conclusion: the total number of spills detected ranged from 126 to 198, and those linked to ships ranged from 29 to 45, suggesting that ship-generated spills remained frequent.[60]

Figure 7.1 graphs data from the U.S. Pollution Incident Reporting System that was established in the early 1970s. For the period between 1973 and 1986, these data show major decreases in incidents attributed to bilge pumping and other intentional discharges from tankers and nontankers, but show those attributed to ballast discharges declining more slowly. Certainly during the period from 1978 to 1981 ballast

60. See United Kingdom Royal Commission on Environmental Pollution, *Eighth Report: Oil Pollution of the Sea* (London: Her Majesty's Stationery Office, 1981), 195.

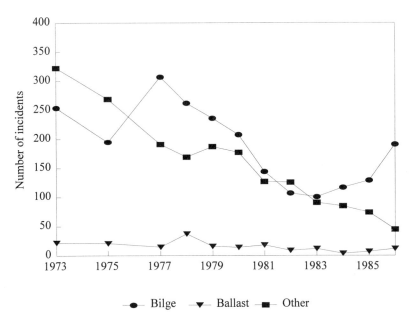

Figure 7.1
Intentional discharges in U.S. waters, 1973–1986
Sources: United States Coast Guard, *Polluting Incidents in and around U.S. Waters* (Washington, D.C.: U.S. Department of Commerce, 1973, 1975–1986).

discharges were no less frequent than they had been in the period imme-diately preceding that.

There are no direct governmental data on the total number of dis-charge violations, because oil-loading states did not enforce those stand-ards. Oil industry surveys conducted during the 1970s to document LOT use, however, providing a reasonable proxy for data on compliance with this standard. During the 1960s oil companies had promoted the use of LOT to avert the use of the implicit equipment regulations of the 1962 amendments, and they claimed that by 1965 78 percent of the fleet had voluntarily adopted LOT.[61] Soon this figure became enshrined as the assumed level of compliance with LOT. Early British estimates assumed that by 1970 80 percent of tankers were using LOT at an efficiency that

61. OP I/21 (15 January 1965), 4; and Pritchard, "Load on Top," 222.

resulted in discharges only double the 1969 legal limit.[62] A 1973 U.S. National Academy of Sciences study also assumed that, although "OILPOL 54/69 was not in force, . . . 80% of oil tankers operated with the load-on-top procedure."[63] During the 1970s the oil companies finally empirically evaluated the use of LOT in their own and independent tankers. Initial surveys in 1972 and 1973 showed that "only one-third of the tankers using their terminals were employing the system well, another third were employing it very poorly, and another third were not using it at all."[64]

Continuation of these surveys—made available after a 1978 congressional request—showed that despite industry efforts, LOT use was neither as widespread nor as effective as the industry had claimed (see figure 7.2).[65] Oil company tankers dramatically reduced their total discharges during the early 1970s. Nevertheless, earlier claims that tankers using LOT would be in compliance were belied by average discharges after 1975 that were three times the legal limit. Although discharges in excess of 1/15,000th of tankers' cargo-carrying capacity constituted OILPOL violations only after 1978, the trend line suggests that many oil company tankers were violating the total discharge limit at least in the first years after 1978.[66]

62. United Kingdom, Programmes Analysis Unit, "The Environmental and Financial Consequences of Oil Pollution from Ships: Report of Study No. VI Submitted by the United Kingdom to the International Marine Pollution Conference 1973," Berkshire, England: Programmes Analysis Unit, 1973, available as MP/Conf/Inf. 14/3 2/Add. 1, and Appendix 1, available as PCMP/2/Add. 1 (12 February 1973), 1.8–1.19.

63. Y. Sasamura, "Oil in the Marine Environment," in *IMAS 90: Marine Technology and the Environment* (London: Institute of Marine Engineers, 1990), 3–4, citing National Academy of Sciences, *Petroleum in the Marine Environment* (Washington, DC: National Academy of Sciences, 1975), 9.

64. M'Gonigle and Zacher, *Pollution, Politics, and International Law*, 110–111. See also Pritchard, "Load on Top," 214. The first results were presented in 1976 by M. P. Holdsworth, "Loading Port Inspection of Cargo Residue Retention by Tankers in Ballast," Paper X/1, in *Report of the Symposium on Prevention of Marine Pollution from Ships Held in Acapulco*.

65. Burton, "Letter," 331.

66. The surveys were discontinued in the late 1970s, and more recent data is not available (Arthur McKenzie, personal communication, 1992).

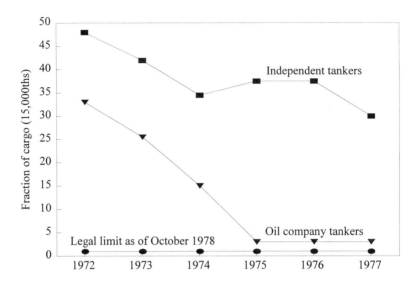

Figure 7.2
Average total discharges, 1972–1977
Source: United States House of Representatives, Committee on Government
Operations, *Oil Tanker Pollution—Hearings: July 18 and 19, 1978* H401–408
(95th Congress, 2nd session) (Washington, D.C.: GPO, 1978), 322.

Not surprisingly, independent tankers did far worse. As an expert
analyst on the subject would later note in testimony to Congress, "In
some ways, one can look back and say that the oil industry was very
stupid in not really forcing tankers, both their own and the independents,
to use LOT efficiently. By not having done so, they then got governments
to force upon them a much more expensive system in the form of
segregated ballast tanks in 1973."[67] By 1977 the two-thirds of the fleet
owned by independents was still discharging 0.2 percent to 0.3 percent
of their cargoes, a level more than thirty times that needed to meet the
requirements that would enter into force the next year. Indeed, these
figures are only slightly less than experts had estimated since the 1960s

67. Mark Zacher, "Testimony," in U.S. House of Representatives, *Oil Tanker
Pollution*, 208.

as typical for tankers practicing no pollution control at all.[68] This information confirms the earlier analysis that independents had fewer incentives to conserve oil than did oil company tankers. It also strongly suggests that violations among the majority of the tankers in the fleet would remain common after 1978, since the discharges already reflected any influence that the 1973 oil price increases might have had on compliance.

Other evidence confirms low LOT usage and hence high total numbers of discharge violations. A 1976 Dutch survey found that only sixteen of seventy tankers were using LOT.[69] Congressional evidence in 1980 showed that "LOT tankers have had difficulty in achieving the required minimum" total discharge required by the 1969 amendments.[70] A 1981 update of the 1973 National Academy of Sciences study revised its estimates, assuming that only 50 percent of the world's tanker fleet was meeting the 1969 discharge limit.[71] The 1989 version of these estimates assumed that 80 to 85 percent of tankers were complying with the total discharge limit, although it provided no evidence of the vast improvement.[72]

Although a tanker using LOT can discharge most slops as part of the next cargo, weather and other factors force even LOT tankers complying with the total discharge limit to discharge slops into port reception facilities. Independent Belgian, Dutch, Norwegian, and West German

68. See, for example, the estimate of 0.3% in James E. Moss, *Character and Control of Sea Pollution by Oil* (Washington, DC: American Petroleum Institute, 1963), 47; and the estimate of 0.4% in OP I/21 (15 January 1965). The growing use of crude oil washing systems may explain the decrease in slops (Gray, "Testimony," 92).

69. MEPC V/Inf. A (27 April 1976).

70. U.S. House of Representatives, Committee on Merchant Marine and Fisheries, *Report to Accompany H.R. 6665, Appendix I, Hearings on Protocol of 1978 Relating to the International Convention for the Prevention of Pollution from Ships, with Annexes and Protocols: August 18, 1980* (96th Congress, 2nd session) (Washington, DC: GPO, 1980), 5.

71. Y. Sasamura, *Petroleum in the Marine Environment: Inputs of Petroleum Hydrocarbon into the Ocean Due to Marine Transportation Activities* (London: Intergovernmental Maritime Consultative Organization, 1981), 12, available as MEPC XVI/Inf. 2 (November 1981).

72. MEPC 30/Inf. 13 (19 September 1990), 15.

studies using different estimating techniques have all concluded that the actual slops received in reception facilities fall dramatically short of the amounts estimated as generated during tanker voyages.[73] A West German survey of tankers found that "half of those subject to control were unable to declare where oil residues had gone."[74] West Germany has also frequently prosecuted cases in which captains did not have "proof of the whereabouts of sludge or oily residues."[75]

More anecdotal evidence supports the general conclusion that violations of discharge provisions remain commonplace. In 1985 West Germany reported still finding "many cases of unlawful discharges."[76] Employees of U.S.-flagged tankers claimed in 1985 that illegally discharging tank cleanings and bilge residues "was the norm on the high seas. In the two years at sea, the ships had never been caught in the act. . . . The informants were adamant that ships 'always' disregard pollution regulation when possible" because of the excessive costs and time lost.[77] Tanker representatives admit that tankers often have to violate discharge limits because of charter arrangements that require them to arrive with clean tanks or because of ports that lack adequate reception facilities.[78] Aerial

73. For the Belgian, Dutch, Norwegian, and German studies, see respectively P. Vanhaecke, "Verontreiniging door Schepen (Ship-source Pollution)," unpublished paper, Antwerp, Belgium, 1990; Marja den Boer et al., "'Loos'-alarm: Afvalolie van de Scheepvaart in de Waddenzee ('Loos'-alarm: Waste Oil from Navigation in the Wadden Sea)," (Groningen, The Netherlands: Werkgroep Eemsmond, 1987); Greenpeace, *Implementation of the Second North Sea Conference: Norway* (Greenpeace Paper 13) (Amsterdam: Greenpeace International, 1989), 24; "Cleaner Oceans: The Role of IMO in the 1990s," *IMO News* 3 (1990), 10; MEPC 26/25 (1988); and MEPC 27/16 (1989), 19.

74. Second International Conference on the Protection of the North Sea, *Quality Status of the North Sea: A Report by the Scientific and Technical Working Group* (London: Her Majesty's Stationery Office, 1987), 14.

75. See, for example, MEPC 30/17, Annex 4 (20 July 1990).

76. MEPC 21/Inf. 8 (21 March 1985).

77. Curtis, "Vessel-source Oil Pollution," 707; and Herbert Carlson, ed., *Quality Status of the North Sea* (Hamburg, Germany: Deutsches Hydrographisches Institut, 1986), 114.

78. See, for example, MEPC 27/5 (17 January 1989); MEPC 27/5/4 (15 February 1989); MEPC 32/10 (15 August 1991); "IMO, Tanker Owners Urge Increase in Facilities Accepting Oily Wastes," *International Environment Reporter,* 8 March 1989, 130; and "Tanker Orders Contribute to Pollution," *International Environment Reporter,* 10 October 1990, 428.

surveillance flights still uncover numerous small, nonaccidental slicks. In 1986 the West Germans observed 186 oil slicks on 240 surveillance flights.[79] That same year the Dutch recorded ninety-five oil slicks on four surveillance flights.[80] The chief of the U.S. Coast Guard's Pollution Prevention and Enforcement Branch claims, "I don't think there's any question that intentional discharges are still occurring out at sea."[81]

This sense that violations remain commonplace has been the foundation for numerous calls to improve enforcement by both treaty members and outside experts.[82] Since violations of any of the discharge standards produce visible slicks, detected oil spills provide a final, if imperfect, proxy for noncompliance. In a 1987 North Sea status report, experts' estimates of illegal discharges ranged from 1,100 to 60,000 metric tons; in 1990 the follow-on review found "no clear or significant trend in observed oil slicks."[83] An extensive meta-analysis of Dutch seabird surveys from 1915 to 1988 showed that "the percentage of seabirds contaminated with oil as well as the total number of oil victims has not changed much over the last few decades [until a recent shift] to a continuous stream of victims caused by chronic oil pollution" and an increase of deaths in the 1980s over the 1970s.[84] In 1990 Yoshio Sasamura, former director of IMO's Marine Environment Division, noted that complete noncompliance with discharge provisions would put 6 million tons of oil into the oceans each year, while complete compliance would put only 0.1 million metric tons into the ocean, but an absence of

79. Second International Conference, *Quality Status of the North Sea,* 14.

80. "Oil Spill Sightings," *Marine Pollution Bulletin* 17 (June 1986), 236.

81. Interview, Commander Lewis J. Beach, U.S. Coast Guard, Washington, DC, 4 April 1992.

82. See, for example, MEPC 27/15/2 (3 February 1989); MEPC 31/5/1 (4 April 1991); and "Cleaner Oceans," 12.

83. North Sea Conference, *1990 Interim Report on the Quality Status of the North Sea* (The Hague: Netherlands Ministry of Transport and Public Works, 1990), 36; and Second International Conference, *Quality Status of the North Sea.*

84. MEPC 29/10/3 (15 January 1990) citing C. J. Camphuysen, *Beached Bird Surveys in the Netherlands 1915–1988: Seabird Mortality in the Southern North Sea Since the Early Days of Oil Pollution* (Amsterdam: Werkgroep Noordzee, 1989).

data led him to "optimistically assume that it would be much nearer to the latter figure at present."[85]

Analysis

Summarizing the empirical evidence, no consistent metrics of actual discharge violations exist that would accurately indicate changes in actual compliance levels over time. Conclusions drawn from a diverse range of proxies, however, provide relative confidence that violations of the content, rate, and total discharge regulations undoubtedly have remained frequent through the 1970s and 1980s, but that discharges probably decreased somewhat between the early 1970s and early 1980s, especially among oil company tankers.

Given the quality of these data, we need to evaluate the effectiveness of the efforts to improve the discharge standards in light of both of these conclusions. The fact that violations of all three discharge standards continue to be frequently detected tends to confirm that the behavior of many tanker operators has remained relatively unchanged. Indeed, some tanker operators appeared to be sufficiently unconcerned about the consequences of detection that between one hundred and two hundred tankers each year conduct blatant enough illegal discharges that their cases are prosecuted by port states or referred to flag states (see table 7.1).

A few pieces of evidence support the argument that a disproportionate share of these discharge violations is committed by independent tankers. First, independent tanker operators had few incentives to adopt the LOT technique that had been promoted by the oil companies in the 1960s, since the benefit of the conserved oil accrued to the cargo owner. In addition, the monitoring and enforcement necessary to deter tanker operators from violating regulations did not actually occur. Since there was no significant improvement in detection, prosecution, and penalization, there are no direct causal pathways to different tanker operator behavior. As the official IMO newsletter stated in 1990, "Little has changed in the three decades since [1962]. The problem is detecting a violation in the first place (which is difficult) and then collecting sufficient

---

85. Sasamura, "Oil in the Marine Environment," 3–6.

evidence to prove the case in court (which has all too often proved to be impossible)."[86] Despite these difficulties, violations of the 60 l/m standard are sometimes prosecuted. In contrast, cases of violations of the 1/15,000th rule are rarely, if ever, detected or prosecuted.[87] Besides, the surveys conducted during the mid-1970s show that independent tankers discharged an average of thirty times the total discharge limit and only slightly less than the average discharges estimated as far back as 1963 for tankers practicing no pollution control. The data even imply that many oil company tankers, while reducing their discharges, probably still exceeded the 1/15,000th standard. Whether most violations are committed by independents or not, a strong case can be made that many tanker operators have not complied with OILPOL and MARPOL's discharge provisions.

Even where the evidence suggests that some tankers decreased their discharges, the treaties were clearly not responsible for the behavioral change. As just noted, no enforcement scheme that could claim responsibility for inducing such a change was ever implemented by port or coastal states. The decrease in average total discharges by oil company tankers from 33/15,000ths in 1972 to 3/15,000ths in 1977 occurred before the 1969 amendments to OILPOL took effect in 1978. Since the amendments called for a change in an operational procedure requiring no lead time to implement, such anticipatory compliance would have been unnecessary. We can more easily and plausibly explain the increased use of LOT by oil company tankers in the 1970s and the absence of a similar increase by independents as due to rising oil prices and the different incentives facing the two groups. The incentives of oil companies to conserve the cargo they owned explain their promotion of LOT in the 1960s, the fact that their average discharges were lower than those of independents in 1972, and their greater decrease in discharges after the 1973 oil price hike. A decrease in discharges most likely occurred because discharges "represent[ed] a loss of valuable refinable oil."[88] All

86. "Cleaner Oceans," 9.

87. Interview, E. J. M. Ball, Director of OCIMF, London, England, 26 June 1991; and interview, James Cowley, Chairman of Britship, Ltd., London, England, 6 July 1991.

88. Resolution A.391(X) (December 1, 1977).

these incentives were even greater for large oil companies because their interests in maintaining their reputations made them more susceptible to general normative and public pressures to comply with the new rules, usually by creating standard operating procedures for all the tankers they controlled.

Negotiators clearly intended for the clean ballast, 60 l/m, and total discharge limits to remedy previous difficulties in enforcing discharge standards. The former made it more likely that surveillance programs could distinguish illegal from legal discharges. The latter made the detection of violations, their linkage to specific ships, and the collection of legally adequate evidence significantly easier. These new primary rules defined compliance in terms of inherently more transparent or "verification suitable" activities.[89] So why did the new rules fail so completely to change the behavior of tanker operators?

At the most fundamental level, the rules failed because they did not correctly analyze and address the sources of noncompliance with the original discharge standards. The incentives that tanker captains had to discharge waste oil at sea were attested to by the fact that it was common practice prior to the 1954 convention. Once the convention delegates settled on the strategy of leaving a tanker's ability to violate unrestricted, they were forced to develop a compliance system based on deterrence. Excluding the normative and social pressures just noted, compliance with the discharge requirements depended on detecting, identifying, prosecuting, and stiffly penalizing those that did violate the rules so that others would not. Although all three changes in the 1969 amendments addressed one or more of these tasks, none of them addressed all the links in the enforcement chain well enough to successfully deter the majority of tanker operators.

Although the clean ballast, 60 l/m, and total discharge standards all made it easier for a captain to know whether discharges were legal, they still had the same incentives to discharge slops at sea as they had had earlier. Despite industry complaints about equipment costs and nonavailability, an incapacity to self-monitor was not the source of noncompli-

89. The phrase is from Wolfgang Fischer, *The Verification of International Conventions on Protection of the Environment and Common Resources* (Julich, Germany: Forschungszentrum Julich GmbH, 1991).

ance. The clean ballast and total discharge standards also made it easier for enforcement authorities to know independently whether observed actions constituted a violation. The discharge process rules (100 ppm, clean ballast and 60 l/m) all required that extensive and expensive efforts be made to detect oil slicks. The total discharge standard could be monitored through far cheaper in-port inspections. However, none of the rules increased the incentives for the responsible states to operate such surveillance and monitoring programs. That this was so proved especially crippling to the total discharge standards, since no actors with the practical ability to monitor discharges had any incentives to take this first step, which was essential to starting the deterrence process.

Credible deterrence required that violations be linked with the responsible tankers. The discharge process rules did not mitigate this problem of "passive-voice violations"—those about which one could only say that a violation had occurred, but not who had perpetrated it—that had plagued the original 1954 standard. Many efforts have been made to make linking spills with tankers easier, including Swedish efforts in the 1970s to "tag" oils with trace elements and more recent "fingerprinting" systems that use the unique characteristics of oil from different oil fields to distinguish the responsible tanker from several suspects.[90] However, both systems require almost immediate sampling from the spill and the suspect tanker.[91] As recently as 1989, IMO adopted a ship identification number scheme so that "pollution offenders do not escape prosecution by a change of identity."[92] The total discharge standard would have completely eliminated this problem, but the absence of monitoring prevented this from happening.

These rule changes did nothing to remedy the historically low rates of successful prosecution and the low fines imposed under the 100 ppm

90. See note 29, chapter 5.

91. Interviews with a Belgian official disclosed that three samples on board the tanker and three from the slick itself must be taken within twelve hours of the spill to satisfy Belgian courts (Ronald Carly, Ministry of Transportation—Marine Division, Brussels, Belgium, 10 June 1991). A French official noted that "judges will not hear a case without samples being taken" (interview, Olivier Laurens, Mission Interministerielle de la Mer, Paris, France, 19 July 1991).

92. "Shipowner Opposes Further Introduction of Pollution Rules Unless Clearly Justified," *International Environment Reporter,* 10 December 1987.

standard. The 60 l/m standard did not address this problem at all. The clean ballast standard theoretically provided a remedy by setting an easy-to-meet standard of evidence, but in practice problems with identification, continued reliance on flag states to prosecute violations outside territorial seas, and each national judiciary's jealous guarding of what constitutes sufficient legal evidence all obstructed the application of the remedy.[93] Despite the clean ballast provisions, states continue to disagree over whether discharges leaving visible traces should automatically be regarded as MARPOL violations.[94] Even when ships are caught obviously violating, the gap between a detected violation and a convicted violation is quite wide.[95] Separate efforts failed to increase the fines states imposed, and the difficulties of collecting fines from extremely mobile tankers remained (see chapter 5). The low prosecution rates and fine levels, especially for the total discharge standards, reflected the absence of incentives for governments, especially flag state governments, to vigorously prosecute and fine those responsible for discharges that caused little individual environmental harm despite the collective damage they caused.

Conclusions

Despite the variations on the discharge prohibition theme represented by the 100 ppm, clean ballast, 60 l/m, and total discharge standards, all four regulated the same actors, and any tanker operator that wanted to discharge waste oil at sea had the same capacity and first-order incentives to do so before the rules took effect as after. The absence of increased first-order incentives to comply therefore forced those drafting the rules to found their compliance systems on a deterrence-based model. Such a

93. IJlstra, "Enforcement of MARPOL," 597; Curtis, "Vessel-source Oil Pollution," 707; and Belgian Marine Environmental Control, "Programme de Surveillance Aerienne de la Pollution Marine," unpublished report, Brussels, Belgium, 1990.

94. See, for example, MEPC 33/4/6 (31 July 1992); MEPC 33/Inf. 28 (31 July 1992); and MEPC 33/20 (5 November 1992).

95. For an interesting account of a blatant discharge violation and the obstacles to prosecution, see A. J. O'Sullivan, "In Flagrante Delicto," *Marine Pollution Bulletin* 2 (December 1971), 180–181.

strategy raises significant obstacles to changing the behavior of actors that have strong economic incentives to continue their current behavior patterns. The efforts to improve compliance were not doomed to failure, however. Rather the new primary rules failed to increase deterrence and thereby increase compliance because none succeeded in properly aligning the practical ability and legal authority necessary to detect and prosecute discharge violations, however they were defined, with strong incentives to do so.

The clean ballast and 60 l/m standards could be monitored by developed countries that had domestic political incentives to undertake aerial and other surveillance programs. But these standards allowed for very little decrease in the resources needed to practically detect most violations or identify their perpetrators. They also left the legal authority for prosecuting and penalizing most violators in the hands of flag states that had shown, and continued to show, little interest in vigorously following up on cases referred by port and coastal states. These standards failed because those actors with incentives to detect and collect evidence of violations could not do so effectively, while those actors that were legally responsible for prosecution had few incentives to prosecute vigorously.

In contrast, the total discharge standards dramatically reduced the resources needed to detect and identify violators and made evidence of prosecution easier to develop. However, improving countries' abilities to conduct enforcement efforts completely ignored the effects that the absence of incentives and legal authority would have on countries making use of these improved rules. Initially only oil companies and flag state governments could inspect tankers in port for total discharge violations, and neither had strong incentives to do so. Oil-loading states had the practical ability to inspect tankers, but under the 1969 amendments to OILPOL lacked the legal right to do so. MARPOL legalized inspections by port states, but ignored the fact that most of these states were not signatories and, like flag states, had few incentives to conduct inspection programs since they lacked domestic environmental pressures. The developed port states that had incentives to inspect lacked the practical ability to do so, since they were not at the end of the ballast voyage.

Although they remain in force today through their incorporation in MARPOL, the rules in the 1969 amendments have continued to have little significant impact on tanker operators' behavior. Many tanker operators have complied with these rules, but their compliance has been due to factors other than the impact of the rules. Theirs has not been the "treaty-induced compliance" described in chapter 2, but rather the "coincidental compliance" that arises from exogenous changes in interests. We can more readily explain most compliance that we have observed as a predictable economic response to rising oil prices in the 1970s, which made the oil being discharged more valuable. These rule changes provide little support for the argument that treaty rules can influence behavior. They do, however, point to many of the obstacles that can prevent them from doing so. Contending that the removal of such obstacles can actually cause treaty-induced compliance demands that we supply evidence from cases in which the rules succeeded in altering behavior.

8
Equipment Standards: A Successful Strategy

By the early 1970s, increasing public concern over the environment and evidence that the load on top (LOT) technique was not being as widely or efficiently used as had been hoped were producing growing dissatisfaction with the effectiveness of the 1954 International Convention for the Prevention of Pollution of the Sea by Oil (OILPOL).[1] The United States pressed for a Marine Pollution Conference in London, both to improve controls on oil pollution and to establish controls on other ship-generated pollutants. To regulate oil pollution, U.S. negotiators particularly sought to move away from discharge standards, which were "in theory . . . a good idea" because they were economically efficient, but were difficult to enforce. Although others had made the same claims in the 1960s regarding the load on top technique, the United States sought the enactment of a ship design standard "because, in essence, it is a self-enforceable mechanism."[2] The United Kingdom supported this view that further reductions in oil pollution now required "eliminating as far as possible this reliance on the human element."[3] Although the negotiations highlighted the fact that requiring equipment would be

1. United Kingdom Royal Commission on Environmental Pollution, *Eighth Report: Oil Pollution of the Sea* (London: Her Majesty's Stationery Office 1981), 192. *International Convention for the Prevention of Pollution of the Sea by Oil,* 12 May 1954, 12 U.S.T. 2989, T.I.A.S. no. 4900, 327 U.N.T.S. 3, reprinted in 1 I.P.E. 332, hereinafter cited as *OILPOL 54.*

2. Compare MP XIII/2(c)/5 (23 May 1972) to J. H. Kirby, "The Clean Seas Code: A Practical Cure of Operational Pollution," in *International Conference on Oil Pollution of the Sea* (Rome, Italy: 1968), 212.

3. MP XIII/2(a)/5 (1 June 1972) and see IMP/CONF/8/16/Add. 1 (3 September 1973). For evidence of British concern regarding enforcement, see G. Victory, "Avoidance of Accidental and Deliberate Pollution," *Coastal Water Pollution:*

costly, "a trade-off was made—a high degree of enforcement at some additional cost for tanker construction."[4] Notably, governments would not have to pay these additional costs.

In a vote of thirty to seven, the negotiators agreed that all large new tankers delivered after 1979 would need to be built with segregated ballast tanks (SBT). SBT reduced the generation of slops by keeping ballast and cargo tanks and lines separate. Besides those of the Americans and British, the "yes" votes included those from reluctant shipping states seeking to avert the adoption of rules requiring the more costly double bottoms proposed by the United States, for which the costs involved were far off due to a recent building boom.[5] Since major oil companies were largely based in the United States, their initial opposition turned into support later as the threat of unilateral U.S. adoption of SBT requirements became more credible. However, governments like France and Japan that represented shipbuilding interests and those representing independent tanker owners—Denmark, West Germany, Greece, Norway, and Sweden—opposed the requirement.[6]

The 1973 International Convention for the Prevention of Pollution from Ships (MARPOL) addressed a range of ship-generated pollutants other than oil, including chemicals, sewage, and garbage. Problems with the chemical regulations made many states reluctant to ratify the convention, and by 1976 "only three minor states had accepted the instru-

---

Pollution of the Sea by Oil Spills (Brussells Belgium: Committee on the Challenges of Modern Society of NATO, 1970), 2.3.

4. The U.S. Treasury Department had found that the costs of SBT far outweighed its benefits, with a benefit-to-cost ratio of 0.0005; see Charles S. Pearson, *International Marine Environmental Policy: The Economic Dimension* (Baltimore: The Johns Hopkins University Press, 1975), 97. On the trade-off between enforcement and cost at the 1973 conference, see Mark Zacher, "Testimony," in U.S. House of Representatives, Committee on Government Operations, *Oil Tanker Pollution–Hearings: July 18 and 19, 1978* (95th Congress, 2nd session) (Washington, DC: GPO, 1978), 210.

5. For an excellent discussion of state positions during both the 1973 and 1978 conferences, see R. Michael M'Gonigle and Mark W. Zacher, *Pollution, Politics, and International Law: Tankers at Sea* (Berkeley: University of California Press, 1979), 107–142.

6. M'Gonigle and Zacher, *Pollution, Politics, and International Law,* 114.

ment."[7] This period of delay, a spate of accidents in late 1976, and continuing enforcement concerns led President Jimmy Carter to propose the 1977 Carter Initiatives, a set of more stringent equipment standards that included requirements for existing tankers to retrofit with SBT. The Tanker Safety and Pollution Prevention Conference (TSPP) was held in 1978. There the United States wanted international agreement if possible, and other states wanted to avert such a "gross overreaction" and to fix other problems with the 1973 agreement.[8]

As at the 1973 conference, enforcement was a major concern. The United States again contended that enforceability was crucial to treaty success, and equipment standards were crucial to enforceability.[9] Other states' positions on the U.S. proposal reflected the fact that retrofits reduced an existing tanker's cargo capacity and the fleet capacity overall by 15 percent. Greece, Norway, and Sweden saw adoption of a retrofit requirement as a way to put many of their laid-up independent tankers back to work. Most states, however, saw SBT as hugely expensive. They proposed crude oil washing (COW) as an environmentally equivalent but cheaper alternative.[10] As a compromise finally accepted by all sides, new tankers had to install both SBT and COW, while existing tankers had to install either SBT or COW.[11] To ensure that these rules took effect

7. M'Gonigle and Zacher, *Pollution, Politics, and International Law*, 122. *International Convention for the Prevention of Pollution from Ships*, 2 November 1973, reprinted in 12 I.L.M. 1319 (1973), 2 I.P.E. 552; and *Protocol of 1978 Relating to the International Convention for the Prevention of Pollution from Ships*, 17 February 1978, reprinted in 17 I.L.M. 1546 (1978), 19 I.P.E. 9451, hereinafter together cited as *MARPOL 73/78*.

8. Sonia Zaide Pritchard, "Load on Top: From the Sublime to the Absurd," *Journal of Maritime Law and Commerce* 9 (January 1978), 194.

9. For a fascinating description of the use of multiattribute analysis in developing U.S. positions for this conference, see Jacob W. Ulvila and Warren D. Snider, "Negotiation of International Oil Tanker Standards: An Application of Multiattribute Value Theory," Appendix A1.1, in Jacob W. Ulvila, *Decisions with Multiple Objectives in Integrative Bargaining*, Ph.D. thesis, Harvard University, Cambridge, MA, 1979.

10. The U.S. contended that it was not equally enforceable, however (Clifton E. Curtis, "Testimony," in House of Representatives, *Oil Tanker Pollution*, 311.)

11. The protocol also applied the equipment requirements to much smaller tankers than had the original 1973 agreement. However, accurate data on compliance by these smaller tankers are not available for analysis.

quickly despite likely ratification delays, both the 1973 convention and the 1978 protocol set different but specific dates by which "new" tankers would need to comply, producing the three equipment standards depicted in table 3.3. The final protocol also removed obstacles to ratification and was made an integral part of the 1973 agreement, and together they are known as MARPOL 73/78.

First-Order Compliance Incentives

These equipment requirements represented a dramatic departure from the strategy underlying the discharge flow and total discharge regulations. The regulations no longer targeted tanker operators directly, but instead sought to prevent discharges by requiring tanker owners to construct and retrofit tankers in ways that eliminated the generation of slops.[12] As both the 1973 and 1978 negotiations illustrated, the costs of SBT and COW were important determinants of negotiation positions and were likely to influence subsequent compliance levels. Although SBT had been developed in the early 1960s and COW by the late 1960s, COW's far greater ability to reduce the waste of oil became especially attractive as oil prices rose in the 1970s.[13] To know whether adoption of these technologies was a response to economic or legal incentives requires us first to evaluate the costs and benefits of the options available to tanker owners.

A tanker owner would have to decide which technology to adopt. For each, costs depended on capital equipment and impacts of the technology on port time and labor costs. Installing SBT on a new tanker involved high costs: high capital and maintenance costs for the additional piping and other equipment and the cost of additional fuel required due to the

12. COW reduces the slops generated by 80% to 85% compared to using no pollution control. See James Cowley, "Regulatory and Environmental Aspects of Marine Pollution," *IMAS 90: Marine Technology and the Environment* (London: Institute of Marine Engineers, 1990), 1–7; and Drewry Shipping Consultants, Ltd., *Tanker Regulations: Enforcement and Effect,* Drewry #135 (London: Drewry Shipping Consultants, Ltd., 1985), 24.

13. On SBT development, see Sonia Zaide Pritchard, *Oil Pollution Control* (London: Croom Helm, 1987), 145. On COW development, see W. M. Kluss,

larger dimensions needed to give the tanker the same cargo-carrying capacity. Retrofits involved these costs as well as an immediate loss of 15 percent of the tanker's cargo-carrying capacity.[14] The benefits of SBT included less time spent in port, both because SBT tankers can ballast and deballast while they load and unload cargo and because they generate fewer slops needing discharge ashore.

These private economics make SBT essentially uneconomic. For new ships, estimates of the capital costs of installing SBT ranged from 2 to 9 percent of total costs and from $180,000 to $7.15 million per ship.[15] Estimates of higher oil transport costs to the United States alone ranged from $10 to $120 million per year.[16] Between 1975 and 1990, the SBT requirement would add up to $2 billion in shipbuilding costs.[17] Even governments advocating retrofits admitted that SBT would increase the cost of carrying oil by 15 percent, and some oil company estimates ran up to 50 percent.[18] As a 1976 Israeli study put it, any owner retrofitting

---

"Prevention of sea Pollution in Normal Tanker Operations," in Peter Hepple, ed., *Pollution Prevention: The Proceedings of the Institute of Petroleum Summer Meeting, 1968* (London: Institute of Petroleum, 1968), 109; M'Gonigle and Zacher, *Pollution, Politics, and International Law,* 262; and Alan B. Sielen and Robert J. McManus, "IMCO and the Politics of Ship Pollution," in David A. Kay and Harold K. Jacobson, eds., *Environmental Protection: The International Dimension* (Totowa, NJ: Allanheld, Osmun & Co., 1983), 160.

14. On new ships SBT can be arranged more optimally to avoid wasting cargo space. This discussion of costs builds extensively on William G. Waters, Trevor D. Heaver, and T. Verrier, *Oil Pollution from Tanker Operations: Causes, Costs, Controls* (Vancouver, BC: Center for Transportation Studies, 1980), 126–129.

15. Philip A. Cummins, et al., "Oil Tanker Pollution Control: Design Criteria vs. Effective Liability Assessment," *Journal of Maritime Law and Commerce* 7 (October 1975), 181–182. See also Pearson, *International Marine Environmental Policy,* 98.

16. Pearson, *International Marine Environmental Policy,* 100. See also Cummins et al., "Oil Tanker Pollution Control," 204.

17. Cummins et al., "Oil Tanker Pollution Control," 181–182. See also Pearson, *International Marine Environmental Policy,* 98.

18. MEPC V/Inf. 4 (8 March 1976), A18. They argued, however that this translated into only a 2% increase in the price of crude oil. The OCIMF estimates are cited in M'Gonigle and Zacher, *Pollution, Politics, and International Law,* 134.

tankers with SBT was investing money "with an anti-economic result."[19] In 1985 the decrease in cargo-carrying capacity was still considered an "overriding disadvantage" of SBT, and experts continue to contend that retrofitting with SBT remains uneconomic.[20] Although particular estimates vary considerably, the costs of SBT have consistently been considered to outweigh its benefits.[21]

In contrast, COW has economic as well as environmental benefits.[22] COW's costs are about one-third those of SBT, entailing the cost of the washing machines and the costs of the additional time and labor needed to wash tanks in port during delivery rather than during the ballast voyage.[23] The COW process has the distinct advantage, however, of increasing the fraction of the cargo delivered, a boon to the cargo owner. The greater amount of cargo delivered—"outturn"—also increases a tanker's effective cargo-carrying capacity and reduces the repair and maintenance costs incurred by the tanker owner by reducing the buildup of sludge.[24]

An exhaustive comparison of the two technologies estimated that the per-voyage costs of COW and SBT together would be $8,000 and those

19. H. N. Wydra, *Potential Economic Effects of the Segregated Ballast Tank (SBT) System on Nations Operating Small Tanker Fleets* (Haifa, Israel: Israel Shipping Research Institute, 1976), 13, available as MEPC V/Inf. 7 (6 September 1976).

20. Drewry, *Tanker Regulations,* 23; and Pieter Bergmeijer, "The International Convention for the Prevention of Pollution from Ships," paper presented at the Pacem in Maribus XVII Conference, Rotterdam, the Netherlands, August 1990, 13.

21. Pearson, *International Marine Environmental Policy,* 97, quotes a U.S. Treasury study cost-benefit ratio of 2000 to 1, while Cummins et al., "Oil Tanker Pollution Control," 172, find the ratio to be 20 to 1. The methods used to account for environmental benefits with these ratios is not clear, however.

22. M. G. Osborne and J. M. Ferguson, "Technology, MARPOL and Tankers: Successes and Failures," *IMAS 90.*

23. Drewry, *Tanker Regulations,* 25.

24. MEPC VIII/Inf. 16 (5 December 1977); Waters, Heaver, and Verrier, *Oil Pollution,* 128; and interview, Captain J. Squires, International Chamber of Shipping, London, England, 4 July 1991.

25. Cost figures are based on Waters, Heaver, and Verrier, *Oil Pollution,* Exhibit VII-4, 124–125. OCIMF estimated that COW saved $10,000 per voyage; see

of SBT alone would be $1,500, while COW alone would *save* $9,000 per voyage.[25] Effective use of LOT, which required no equipment, was an even cheaper way to reap the benefits of reduced oil waste, saving $17,000 per voyage. Both tanker and cargo owners, whether oil companies or independents, preferred LOT, then COW alone, SBT alone, and COW and SBT together in that order. The pressures during the 1970s for more stringent environmental protection overrode these preferences, however, and MARPOL required "the most costly, but not the most effective of the control technologies."[26] The compliance system had to counteract these strong economic disincentives if compliance with these standards was to be any different than with the discharge standards.

Expected Pathways of Influence

The equipment standards of MARPOL 73/78 relied on preventing rather than punishing violators to induce compliance. They depended far more on coercing tanker operators to comply in the first place than on deterring them from committing violations as the discharge standards did. To implement such a strategy required a far greater focus on "premonitory" control measures, i.e., efforts to inspect and survey behavior before violations were committed, than on detecting and investigating them afterward.[27] Several elements of the compliance system were expected to ensure compliance.

First, unlike discharging oil, buying a tanker requires the knowledge and consent of other actors. Unlike illegally discharging oil, which can readily be done with only a captain and crew involved, building a tanker is done only on order; someone seeking to order a tanker without the required equipment would need cooperation in committing an illegal act from at least three other actors: a builder, a classification society, and an insurance company. Tanker purchases were already subjected to exten-

---

MEPC VIII/Inf. 16 (5 December 1977). Exxon estimated that COW would cost $310 per ton of oil saved, but recommended it over the alternative of retrofitting tankers with SBT, which would cost $4,210 per ton of oil saved (Exxon Corporation, "Crude Tanker Pollution Abatement," position paper, April 1976).

26. Waters, Heaver, and Verrier, *Oil Pollution*, 129.

27. For the distinction between "coerced compliance" regulatory systems and "deterrent" systems, as well as premonitory vs. postmonitory control, see Albert

sive private and governmental monitoring for safety, financing, and insurance purposes. Although classification societies and insurance companies face stiff price competition, they do not compete by making certification standards more lax. They regularly update classification standards to incorporate all international legal construction requirements. Classification societies have shipyard representatives who "play a vital role in monitoring compliance" during the construction process.[28] Since acquiring insurance depends on a tanker's being classified, noncompliance with equipment standards would effectively preclude the purchase of insurance coverage. In short, by regulating an interpersonal transaction rather than an autonomous action, the new rules allowed for the identification of potential violators and made it harder to actually commit a violation.

MARPOL also required flag state governments or classification societies nominated by them to survey all tankers to ensure compliance before issuing the required International Oil Pollution Prevention (IOPP) certificate and to conduct periodic inspections thereafter.[29] Although developed states concerned about pollution could be expected to ensure that noncompliant tankers did not receive certificates, many developing flag states have traditionally turned over such inspections to classification societies to avoid the administrative costs involved. These flag states might have applied international standards less rigorously than other states had they been responsible for the surveys. However, they could delegate this task to a relatively few classification societies, most of which are large international corporations whose revenues depend on their public reputations. Thus they had stronger incentives, as well as the ability and authority, to withhold certificates from tankers that did not meet international standards. MARPOL also required that flag states or the surveyors nominated by them ensure that corrective action was taken if a tanker posed a threat to the environment, detaining it if necessary.[30]

J. Reiss, Jr., "Consequences of Compliance and Deterrence Models of Law Enforcement for the Exercise of Police Discretion," *Law and Contemporary Problems* 47 (Fall 1984).

28. Interview, John Foxwell, Shell International Marine, London, England, 27 June 1991.

29. Annex I, Regulations 4 and 5, *MARPOL 73/78*.

30. Article 5(2) and Annex I, Regulation 4(3)(d), *MARPOL 73/78*.

These first two features of the compliance system presented large obstacles to a tanker owner's being able to violate the equipment standards despite having the incentives to do so. Classification societies, insurance companies, and flag state inspectors could withhold the papers necessary for a tanker owner to conduct business in international oil markets, thereby frustrating any tanker owner's attempt to reap the benefits of sidestepping these standards. Even if one could buy a noncompliant tanker, numerous other individuals and organizations had the ability and incentives to prevent others from trading with it on world markets.

To the extent these elements of the compliance system failed, MARPOL provided methods that would identify and sanction the owner of any tanker that was built without proper equipment. By its very nature, such an equipment violation would be transparent throughout a tanker's lifetime to any state conducting inspections. To ensure this, MARPOL explicitly gave port states the right to inspect a tanker's equipment if the tanker was suspected of not conforming with the standards for the issuance of an IOPP certificate. This provision not only checked violations by tanker owners, but also identified (and hence deterred) any classification societies or flag states that were issuing false IOPP certificates.

Equally important, port states were obligated to detain any tanker that had a false certificate or inadequate equipment "until it can proceed to sea without presenting an unreasonable threat of harm to the marine environment."[31] MARPOL also reminded governments of their preexisting right to deny noncompliant tankers entry to their ports.[32] In essence, this meant that a tanker lacking an IOPP certificate or not built in compliance with the SBT and COW requirements could be barred from doing business and even impounded. The administrative sanctions of barring and detaining effectively circumvented the legal and evidentiary problems that had made even clear violations of discharge standards difficult to successfully prosecute. Port state authorities could detain a ship until it was retrofitted with SBT or COW. Compared to fines that have consistently averaged below $10,000 for discharge violations,

31. Article 5(2), *MARPOL 73/78*.
32. Article 5(3), *MARPOL 73/78*.

retrofitting could cost millions of dollars. As was noted in chapter 5, several major oil-importing states conduct port state inspections, and a few have detained tankers.[33] The huge opportunity costs of having a ship barred from port or detained would force a tanker owner to think twice before deciding to order or retrofit a tanker that would not fully comply with MARPOL's equipment regulations.

As experience with discharge standards had shown, establishing enforcement requirements was not necessarily equivalent to establishing actual enforcement infrastructures. Inspection, surveillance, and monitoring seemed far more likely for equipment standards than for discharge standards precisely because they did not require new, dedicated infrastructures, but instead piggybacked on existing ones. Classification societies and insurance companies already inspected tankers; adding inspections for pollution prevention equipment required only minor modifications to existing procedures and minimal additional effort. Most governments already had some form of in-port inspections to address customs and safety concerns, and many specifically added IOPP certificates and MARPOL-required equipment to their inspection criteria. As was noted in chapter 5, the United States, Japan, Canada, the fourteen European states of the Memorandum of Understanding (MOU), Poland, Russia, and even a group of Latin American states all have port state control efforts that include pollution inspection procedures.

Although many nations were inspecting IOPP certificates, the value of such inspections depended on high-quality initial surveys by the flag states or classification societies. Data from both IMO and MOU states corroborate that IOPP discrepancy rates of 9 to 11 percent of ships inspected in the first year after MARPOL took effect quickly declined to 1 to 3 percent by 1989 (see table 5.3).[34] Although they combine tankers having no certificates with those having inaccurate certificates, these low recent figures suggest that, after an initial period of learning how to issue

33. The United States, Japan, Canada, and the European Memorandum of Understanding states all practice extensive port state control procedures.

34. The IMO and MOU data include both tankers and nontankers in the "number of ships inspected" figures. Given that tankers make up approximately seven thousand of the seventy-five thousand ships currently in the world fleet and that IMO and MOU inspections cover twenty-five thousand and ten thousand ships, respectively, these data probably provide a major sample of the tanker fleet.

and inspect IOPP certificates, almost all tankers now receive thorough and accurate IOPP surveys. Like port states, flag states and classification societies appear to have become active participants in the equipment subregime's compliance information system.

Although many states have adopted procedures to inspect tankers and their IOPP certificates, states have proved more reluctant to detain non-compliant ships. Evidence from states that are detaining tankers shows that there are many substandard tankers on the international market. Yet only a few states have frequently used detention. States treat MARPOL's detention clause, despite its phrasing, as a right, not an obligation. The advantage of detention lay in the fact that even its low level of usage by a few major oil-importing states posed major risks to tanker owners. Even a low probability of being detained or barred from certain oil markets was a potent threat with which tanker owners had to reckon. Although few states detain ships, those that do control a significant (though not a majority) share of the oil market. When the threat of detention was coupled with the obstacles to getting a tanker without SBT or COW built or retrofitted, classified, and insured in the first place, tanker owners faced a much greater likelihood that noncompliance would be detected and that, if sanctions were imposed, they would prove extremely costly.

Observed Compliance by Tanker Owners

The likelihood of being prevented from or detected in violating the SBT requirement and the costs of being detained or prevented from trading provided strong incentives to comply. These second-order incentives to install SBT on tankers arose solely as a result of the MARPOL agreement. The need for an IOPP certificate, the attendant surveys by classification societies, the inspections by many port states, and the detentions and barring from port by a few states involved an international system founded on requirements established by MARPOL. To say that the monitoring and enforcement created strong pressures for compliance is not, however, to say that compliance resulted or that, if it did, it was due to the treaty. The high costs of SBT relative to other means, including COW, of reducing oil waste create strong incentives for tanker owners to violate the SBT requirement. These competing pressures made it hard to forecast compliance levels. Indeed, in the same year that one analyst claimed that the enormous costs of SBT would make compliance "negligible,"[35] another claimed that "all new tankers are now being built to these 1973 specifications."[36] Therefore, it becomes crucial to empirically evaluate actual compliance levels.

MARPOL 73/78 established clear requirements for large crude oil tankers: those delivered after 1 June 1982 must have SBT and COW, those delivered after 31 December 1979 but before 1 June 1982 must have SBT alone, and those delivered before 31 December 1979 must have been retrofitted with SBT or COW before 1 October 1985.[37] Although the major pathway by which these regulations were to influence tanker owners' behavior was by preventing violations in the first place, finding direct evidence of such influence proves inherently difficult. There is no evidence of how many companies attempted to buy new tankers without SBT and COW but were refused by builders or classification societies. We are even less likely to know how many were deterred from such attempts by the fear that doing so would both prove unsuccessful and tarnish their reputations, making them appear to be seeking to avoid the law. Likewise the captains of tankers coming in for repairs and maintenance undoubtedly found it difficult to explain why they were not plan-

35. Charles Odidi Okidi, *Regional Control of Ocean Pollution: Legal and Institutional Problems and Prospects* (Alphen aan den Rijn, the Netherlands: Sijthoff and Noordhoff, 1978), 34.

36. Zacher, "Testimony," 208.

37. 1 October 1985 was two years after MARPOL entered into force. Large tankers include all tankers over seventy thousand deadweight tons; see IMO, *Regulations for the Prevention of Pollution by Oil* (London: IMO, 1985), 39. Although MARPOL imposes requirements on small tankers, the compliance figures available for smaller tankers include both tankers built to comply and those "designated" into compliance. This is because, although MARPOL requires new product carriers to install SBT only and new crude carriers to install SBT and COW, the "Unified Interpretation" of MARPOL requires that the IOPP certificate designate tankers as crude or product oil carriers based on their installed equipment; see IMO, *Regulations for the Prevention of Pollution by Oil* (Including unified interpretation of the provisions of Annex I) (London: IMO, 1985). Thus, although the equipment required depends on tanker type, tanker type depends on the equipment on board. In contrast, the compliance of large tankers need only be measured against the crude carrier requirements, since product carriers rarely exceed seventy thousand deadweight tons (interview, Sean Connaughton, American Petroleum Institute, Washington, DC, 4 April 1992).

ning to retrofit with SBT or COW as required. As the low levels of noncompliance described below reveal, however, these obstacles to committing equipment violations most likely have played a major role in deterring violations.

Fortunately the greater ease of monitoring equipment standards than discharge standards also allows for the development of more consistent and comprehensive data on compliance levels. Several private companies have extensive and detailed data bases on tankers, including their equipment, to help charterers make decisions about what tankers to use to carry their oil. Before MARPOL took effect but after its equipment requirements had been adopted, a 1981 study already began to show that tanker operators were installing COW and SBT as required by MARPOL.[38] The sharp increase in the number of tankers constructed with SBT, from 8 to 27 percent of all tankers, suggests that many tanker operators were already complying with MARPOL's requirements despite the costs involved. Indeed the 27 percent figure significantly underestimates compliance levels since SBT were required only for tankers built in the last two years of the period from 1976 to 1981. The pre-1976 data show that tanker operators were certainly not installing SBT before MARPOL required it. At the same time, 26 percent and 32 percent, respectively, of tankers in the two groups had already installed COW despite the fact that those built in 1980 and 1981 would never be required to have it and those built before 1980 would not have to install SBT or COW until 1985. The data also confirm that adding COW was more common than adding SBT when retrofitting tankers and that owners rarely installed SBT alone.

Similar data from Clarkson Research Studies on the tanker fleet at the end of 1991 provide far clearer evidence of compliance levels under MARPOL's equipment requirements.[39] Table 8.1 delineates the on board equipment of the current tanker fleet. The table breaks out the total number of tankers into the different year groups established by MARPOL's

38. Drewry Shipping Consultants, Ltd., *The Impact of New Tanker Regulations,* Drewry no. 94 (London: Drewry Shipping Consultants, Ltd., 1981).

39. Electronic version of Clarkson Research Studies, Ltd., *The Tanker Register* (London: Clarkson Research Studies, Ltd., 1991) provided to author by Clarkson Research Studies, Ltd.

Table 8.1
Percentage of tankers equipped with SBT and/or COW
(crude oil tankers over 70,000 deadweight tons)[a]

Equipment on board	Tanker construction date		
	1979 and prior	1980–82	Post-1982
SBT and COW	32	94	98
SBT or COW	94	99	100
Total SBT (alone + with COW)	36	98	99
Total COW (alone + with SBT)	89	95	99
SBT alone	4	4	1
COW alone	58	1	1
Neither SBT nor COW	6	1	0
MARPOL requirement	SBT or COW	SBT only	SBT and COW
Compliance level (%)	94	98	98

Source: Electronic version of Clarkson Research Studies, Ltd., *The Tanker Register* (London: Clarkson Research Studies, Ltd., 1991), provided by Clarkson Research Studies, Ltd.

1973 and 1978 standards, documenting that the level of compliance with the MARPOL equipment requirements has been extraordinarily high. Some 94 percent of tankers built in 1979 or earlier have installed SBT or COW, 98 percent of those built in 1980 through 1982 have installed SBT, and 98 percent of those built after June 1982 have installed both. Striking progress was made in ten years: although only 45 percent of tankers built from 1976 to 1981 and only 26 percent of pre-1976 tankers had SBT or COW in 1981, 94 percent of older tankers had installed COW or SBT by 1991 and all post-1982 tankers had done so. Owners scrapped tankers that had neither SBT nor COW far more frequently in the early 1980s than they scrapped compliant tankers, further supporting the claim that equipment requirements were driving behavior.[40] Tankers from all countries have complied, whether or not those countries supported equipment requirements during negotiations.

40. Drewry, *Tanker Regulations*, 21–22.

The data not only document compliance levels, but also confirm the sources of compliance suggested by the 1981 data and the incentives identified earlier. Owners have generally not installed SBT except when MARPOL required it. MARPOL requires SBT (alone or with COW) on all tankers built after 1979, and 98 percent and 99 percent of tankers installed SBTs in the 1980–1982 and post-1982 periods, respectively. In contrast, owners that could choose to retrofit their pre-1980 tankers with SBT or COW put SBT on only 36 percent of tankers, but put COW on 89 percent. The fact that only 4 percent installed SBT alone while 58 percent installed COW alone suggests that owners installed SBT only when a tanker was already in dock to be retrofitted with COW, when the marginal cost of SBT installation was very low. Tanker owners also installed COW on 95 percent of large tankers built between 1980 and 1982, even though only SBT was required. COW's economic benefits, rather than MARPOL, appear to be the reason for COW installation. Figure 8.1 depicts tankers in the fleet in 1991 by year of construction. This figure clearly illustrates the much greater first-order incentives tanker owners had to adopt COW rather than SBTs. In every year prior to 1980, at least 70 percent—more often 85 to 90 percent—of the new tankers built had COW installed. In contrast, never more than 60 percent—and often fewer than 30 percent—installed SBT. The dramatic increase in the percentage of new tankers built with SBT that began in 1977 preceded the 1980 MARPOL requirement, but only by the few years needed to fit new construction into shipyard scheduling requirements.

The IMO and MOU statistics on IOPP certificate discrepancies provide additional support for the finding that compliance rates have been quite high. Neither source distinguishes between major IOPP discrepancies like the absence of SBT or COW and the more likely problem of incorrect paperwork. IOPP certificate deficiencies were found for 9 to 11 percent of tankers in 1984 (see table 5.3), but the percentage has since dropped back to 1 to 3 percent. Many of these discrepancies probably involve technical deficiencies or paperwork problems rather than actual failures to have SBT or COW on board. Nevertheless, the figures correlate well with the very low violation rates just delineated and suggest that, after an initial period of adjustment, violations of equipment standards and the supporting IOPP certificate requirements have become quite uncommon.

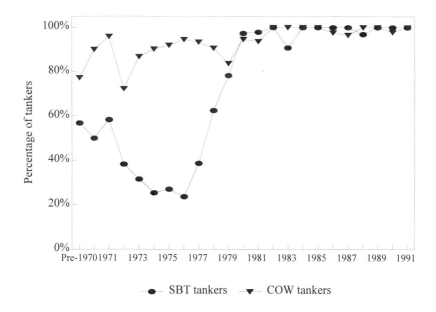

Figure 8.1
Percentage of tankers equipped with SBT and COW
(1991 fleet by year of construction)
Source: Electronic version of Clarkson Research Studies, Ltd., *The Tanker Register* (London: Clarkson Research Studies, Ltd., 1991), provided by Clarkson Research Studies, Ltd.

Other anecdotal evidence reinforces these conclusions. A 1981 study found that tanker owners were anticipating the upcoming MARPOL requirements and "retrofitting their fleets at a fairly rapid pace. . . . Virtually every day there are additional reported retrofits [of SBT and COW] to existing vessels."[41] A 1987 conference of marine pollution experts attributed the decreased discharges to "the fitting of new tankers with segregated ballast tanks."[42] A U.S. government analysis in 1989 assumed that all new tankers flagged in MARPOL signatory states complied with the SBT standard.[43] In 1990 the Dutch delegate to the Marine

41. Drewry, *The Impact of New Tanker Regulations*, 25.

42. Second International Conference on the Protection of the North Sea, *Quality Status of the North Sea: A Report by the Scientific and Technical Working Group* (London: Her Majesty's Stationery Office, 1987), 57.

43. MEPC 30/Inf. 13 (19 September 1990), 8.

Environment Protection Committee of IMO wrote that "most tankers built since 1973 have conformed to many features of the original MARPOL Convention—for example, SBT."[44] A representative of the American Bureau of Shipping estimated that "all new tankers are built with SBT and COW."[45] Almost all experts in the field assume that all tankers comply with the equipment standards applicable to their age and size.[46]

Although all these analyses link adoption of SBT to the MARPOL rules, there is considerable evidence that COW's adoption preceded, and hence was not exclusively due to, international legal requirements. COW requirements were incorporated into MARPOL only in 1978 and did not apply to new tankers until 1982. Yet by the mid-1970s oil companies had adopted COW as "standard operational procedure [for] improving outturn, reducing sludge deposits and thus maintaining cargo dead-weight."[47] Three years before the 1978 conference, British Petroleum and Exxon had both adopted COW in all their very large tankers.[48] Experts explained a decrease in tanker slops after 1975 as "almost certainly attributable to the introduction of crude-oil washing."[49] Although some companies did install COW, many still did not, and a 1976 Dutch survey found that only two of seventy tankers were using COW.[50] COW installations increased into the 1980s "both to improve oil cargo outturn as an economic incentive and to implement the pending MARPOL 73/78 requirements."[51] The 1989 U.S. government study assumed that existing

44. Bergmeijer, "The International Convention for the Prevention of Pollution from Ships," 12.

45. Telephone interview, Jim Wolling, American Bureau of Shipping, 10 February 1992.

46. For example, E. J. M. Ball of OCIMF claimed that the SBT and COW requirements have all been "perfectly complied with" (interview, London, England, 26 June 1991).

47. Osborne and Ferguson, "Technology, MARPOL and Tankers," 6-2.

48. Waters, Heaver, and Verrier, *Oil Pollution from Tanker Operations*, 95. See also William Gray, "Testimony," in House of Representatives, *Oil Tanker Pollution*, 12.

49. Gray, "Testimony," 92.

50. MEPC V/Inf. A (27 April 1976).

51. Y. Sasamura, *Petroleum in the Marine Environment: Inputs of Petroleum Hydrocarbon into the Ocean Due to Marine Transportation Activities* (London:

tankers with the choice of COW or SBT installed COW.[52] Although none of these analysts have had access to the detailed empirical data provided here, they have been unanimous in concluding that all tanker owners comply with MARPOL's equipment requirements but seek to avoid the expense of SBT when possible.

In summary, both hard empirical data and expert opinion substantiate the conclusion that the level of compliance with MARPOL's requirements has been impressively high despite the high costs involved. More important, tanker owners' installation of SBT and COW equipment neatly follows a model that predicts that tanker owners' decisions will reflect both independent economic incentives and MARPOL's regulatory pressures. Tankers built before 1980, which can comply via the installation of either COW or SBT, predominantly install COW because it provides greater economic benefits than SBT. Tankers built between 1980 and 1982, which must install SBT have done so, but have also installed COW even though it is not required. Tankers built after 1982 must install both SBT and COW and have installed both.

Analysis

The levels of compliance with the SBT and COW equipment standards provide a striking contrast with the levels of compliance with the discharge standards analyzed in the previous chapter. For the discharge standards, treaty rules could be quickly discounted as the reason for behavioral change since, where there was evidence that behavioral change had occurred, economic factors provided simple and convincing explanations. In contrast, especially in the case of SBT, the compliance system created by the equipment standards and supporting regulations appears to have brought about much of the compliance. Before we can

---

Intergovernmental Maritime Consultative Organization, 1981), 11. Although SBT was already required under MARPOL rules, no mention is made of SBT compliance rates.

52. MEPC 30/Inf. 13 (19 September 1990), 8. This analysis was affirmed by Jim Wolling of the American Bureau of Shipping in a telephone interview on 10 February 1992: "Existing crude oil carriers that must install either SBT or COW tend to install COW."

claim that the radical change in tanker owners' behavior constituted "treaty-induced compliance" in the sense described in chapter 2, however, we need to evaluate several rival hypotheses that identify other reasons for compliance. Three major alternative explanations present themselves: first, economic forces, not international rules, explain the fleet's adoption of SBT; second, U.S. interests and power would have caused the same behavioral change even without international agreement; and third, growing public concern about oil pollution would have led owners to adopt these technologies even without MARPOL.

Several types of evidence discount the economic hypothesis. The only economic benefit of SBT was that it reduced the amount of oil wasted. Therefore SBT adoption, as a long-term investment, should fluctuate with current or forecast oil prices. Yet, contrary to expectation, SBT has been installed on new tankers almost exclusively since 1979, the year in which oil prices peaked and began a steady decrease. In addition, the absence of any detectable change in the number of violations of the discharge standards by independents between the 1970s and 1980s suggests that exogenous changes in the world oil market did not create new incentives to install SBT. During this period, the fleets of both independent and oil company tankers underwent major overhauls in which all new tankers and many older ones installed technologies that significantly reduced the amount of waste oil generated. If economic forces caused such a radical change, they certainly should also have evidenced themselves in decreased numbers of discharge violations.

As has already been noted, for tanker owners that wanted to reduce the amount of oil wasted, SBT provided the least attractive alternative. Economic calculations, industry negotiating positions, and tanker owners' behavior all confirm this. SBT has consistently been evaluated as "anti-economic." LOT and COW both provided positive returns on investments.[53] Yet as late as 1991, oil and shipping interests opposed mandatory SBT retrofits as too expensive, a stance inconsistent with SBT's being economic.[54] Although SBT has been available since the 1960s, very few new tankers installed SBT until they were required in

53. Drewry, *Tanker Regulations*, 21.

54. SBT retrofitting was reintroduced to reduce accidental pollution; see ICS, OCIMF, and INTERTANKO, *Oil Tanker Design and Pollution Prevention* (Lon-

the late 1970s. Since then the different retrofit rates among tankers that have had to retrofit with either SBT or COW have confirmed that tanker owners have not seen SBT as an economically attractive alternative: of pre-1979 tankers currently in the fleet, only 36 percent have installed SBT, while 89 percent have installed COW. As representatives of Shell and Lloyd's noted in 1990, "Left to himself, no owner will, understandably, wish to be placed at a commercial disadvantage to his competitors by introducing segregated ballast on his ships if the whole industry is not doing likewise."[55] Put more bluntly, "If there were not a regulatory requirement, there would not be SBT."[56]

In contrast to SBT, economic factors can explain some of the COW installations observed. The costs and benefits provided a "strong economic incentive for crude tankers to have COW."[57] Although it had been developed in the 1960s, "only when its use became profitable after the OPEC price rise was the system touted for its environmental advantages."[58] Oil price increases in both 1973 and 1978 would have increased the incentives to adopt COW. Oil companies were already installing COW eight years before MARPOL required it. By 1981 a quarter of tankers more than five years old had already retrofitted, and by 1991 more than 90 percent had done so. Indeed, at the 1978 protocol conference oil companies proposed COW as an environmentally equivalent alternative to avert requirements to retrofit all tankers with SBT while simultaneously forcing independents to install COW, thereby increasing outturn of the cargoes they owned.

Although economic factors explain the COW installations of some tankers, it seems unlikely that 90 percent of old tankers and 99 percent of new tankers would have COW without MARPOL. LOT offered a far cheaper means of increasing outturn and represented greater future

---

don: March, 1991), available as MEPC 31/8/5 (4 April 1991). Bergmeijer states that it remains "uneconomic to convert some existing tankers" to SBT; see "The International Convention for the Prevention of Pollution from Ships," 13.

55. Osborne and Ferguson, "Technology, MARPOL and Tankers," 6-2.

56. Interview, Sean Connaughton, American Petroleum Institute, Washington, DC, 8 April 1992.

57. Drewry, *Tanker Regulations,* 21.

58. M'Gonigle and Zacher, *Pollution, Politics, and International Law,* 262.

flexibility than COW, and thus should have been the procedure of choice.[59] In addition, although oil companies owning their own tankers could balance their costs as tanker owners against their benefits as cargo owners, independent tanker owners might well have avoided or deferred COW installation.[60] Independent tanker owners had to pay the direct and immediate capital costs of installing COW. Cargo owners reaped the immediate benefits of increased outturn, covering only those amortized COW costs that market conditions allowed independents to reflect in transportation costs. Gluts in the tanker market often force independents to operate for long periods of time at costs that reflect only variable operating costs and do not include returns on their capital investments. Facing long investment recovery periods, the two-thirds of the market composed of independents had few incentives to install COW. If oil company tankers would have installed COW even in the absence of MARPOL, it seems unlikely that almost all independents would have done so. More plausible is the conclusion that a significant fraction of tankers that installed COW, and almost all tankers that installed SBT, were not led to do so by economic pressures.

To claim that tankers adopted SBT because of treaty requirements demands that we demonstrate that it was not the domestic regulations of a dominant state that caused this change. Indeed some analysts have claimed that U.S. hegemony explains the high MARPOL compliance rates.[61] Certainly the U.S. threats in 1972 and 1977 to impose SBT requirements on U.S. ships and all foreign ships entering U.S. ports motivated the passage of the SBT requirements in MARPOL and its protocol.[62] Admitting that MARPOL's rules reflect hegemonic pressures,

59. See Waters, Heaver, and Verrier, *Oil Pollution from Tanker Operations.*

60. See Drewry, *Tanker Regulations.* In the glutted oil tanker market of the 1970s, most companies counted themselves lucky if they covered variable costs, let alone fixed costs. Although they could absorb, over the short term, COW installation costs, the long-term outlook would certainly have made some independents reluctant to install COW.

61. Jesper Grolin, "Environmental Hegemony, Maritime Community, and the Problem of Oil Tanker Pollution," in Michael A. Morris, ed., *North-South Perspectives on Marine Policy* (Boulder: Westview Press, 1988).

62. The U.S. later followed through on such threats by requiring tankers smaller than MARPOL required to retrofit with SBT or COW and set a faster timetable (Drewry, *Tanker Regulations,* 11).

Table 8.2
Gross tonnage of tankers launched as a percentage of the world total[a]
(countries with over 1% of the world total gross tonnage in four or more periods)

Country of build	1955	1960	1965	1970	1976	1980	1985	1990
Denmark	2.4	3.4	1.9	3.6	3.7		5.6	3.7
France	8.9	9.5	4.5	5.7	4.5			
Germany	10.0	11.0	5.1	6.2	4.9	1.1		
Italy	3.3	8.3	4.4	2.4	2.4			
Japan	16.2	16.3	50.6	50.7	39.9	72.3	48.8	51.3
Korea					2.8	4.2	24.0	18.6
Norway	3.9		2.5	3.2	2.6	2.2		
Spain	1.0	1.9	1.7	4.7	7.4	3.0		5.8
Sweden	13.2	11.9	16.3	13.0	13.2	3.7	3.4	
United Kingdom	26.6	16.9	7.8	3.7	4.8			
United States	2.1	7.4		2.0	4.2	3.2	2.0	
Yugoslavia		1.2	1.8		1.3		3.5	7.5
Other	0.5	4.0	3.4	2.3	2.7	3.2	4.3	3.7
Total world tonnage (millions of dead-weight tons)	2,437	3,649	5,393	10,031	15,420	4,716	2,928	4,686

Source: Lloyd's Register of Shipping, *Annual Summary of Merchant Ships Completed* (London: Lloyd's Register of Shipping, various years).
[a]Includes all oil tankers over 100 gross registered tons. No entry signifies that the country launched less than 1% of the world gross tonnage in that year.

however, does not imply that subsequent behaviors resulted from that hegemonic pressure. Resources adequate to convince diplomats to vote for certain resolutions during a conference may prove inadequate to bring about corresponding changes in economic behavior. Available evidence suggests that the United States could not have induced so many tanker owners to install SBT using unilateral measures.

Although the United States wields tremendous diplomatic leverage, it wields nothing near hegemonic power in world oil markets. The United States could influence tanker construction through direct control of construction, through control of tanker registries, or through control of imports. Yet it lacks hegemonic power in all of these realms. The U.S. share of new tankers launched has never exceeded 5 percent of the world total since the United States became concerned about oil pollution in the late 1960s (see table 8.2). Indeed since then Japan, which resisted SBT requirements at both the 1973 and 1978 conferences, has controlled 40

Table 8.3

World tanker registries—percentage
(countries averaging more than 1% of the world total gross tonnage)

Country	1955	1960	1965	1970	1975	1980	1985	1990
Denmark	1.9	2.1	1.8	1.6	1.4	1.6	1.6	1.5
France	4.4	4.5	4.2	4.0	4.6	4.4	3.1	1.3
Germany	1.1	1.4	1.6	1.9	1.8	1.6	1.0	0.2
Greece	0.7	2.2	3.1	4.5	5.5	6.7	6.8	5.8
Italy	4.6	4.2	3.6	3.2	2.7	2.7	2.6	1.9
Japan	2.8	3.8	6.9	10.7	11.7	10.1	10.2	5.6
Liberia	8.9	17.3	19.3	22.4	27.7	28.5	22.8	21.3
Netherlands	3.4	3.3	2.8	2.3	1.8	1.4	0.5	0.4
Norway	15.8	15.4	15.2	10.3	8.9	6.9	5.3	8.0
Panama	8.1	6.0	4.4	3.8	3.7	3.9	6.1	7.5
Soviet Union	0.8	1.7	3.9	4.0	2.5	2.7	3.3	3.1
Spain	0.8	1.1	1.1	1.7	1.7	2.8	2.1	1.1
Sweden	3.4	3.2	2.5	1.9	2.0	1.1	0.7	0.4
United Kingdom	19.9	17.0	14.4	14.0	10.7	7.6	4.3	1.8
United States	16.3	11.2	8.2	5.4	3.4	4.5	5.4	6.3
Other	7.2	5.8	7.1	8.3	9.8	13.6	24.3	33.8

Sources: Michael M'Gonigle and Mark Zacher, *Pollution, Politics and International Law: Tankers at Sea* (Berkeley: University of California Press, 1979), 56–57; and Lloyd's Register of Shipping, *Statistical Tables* (London: Lloyd's Register of Shipping, various years).

Table 8.4

Percentage of total world oil imports by country
(countries with over 1% of imports in all years)

Country	1959	1965	1970	1975	1980	1985
Brazil	1.8	1.6	1.4	2.5	2.9	2.6
France	9.2	12.3	11.0	10.2	8.7	6.6
Germany (FRG)	4.6	4.8	3.6	2.7	2.6	1.9
Italy	7.8	11.0	12.8	9.2	7.9	8.6
Japan	6.0	11.2	15.5	16.4	14.2	18.3
Neth Antilles	12.3	6.6	4.3	1.8	4.4	0.9
Netherlands	4.9	6.2	8.5	8.0	6.7	6.7
Spain	1.3	1.3	2.2	3.0	3.3	4.0
United Kingdom	12.5	10.4	9.4	6.4	3.0	2.5
United States	19.1	10.7	6.1	13.9	16.8	13.2
Other	20.5	23.9	25.2	25.9	29.5	34.7

Source: United Nations, *Statistical Yearbook* (New York: United Nations, various years).

percent to 75 percent of all new construction.[63] The United States could not have caused the observed change in tanker construction through control of the shipyard.

The number of tankers registered in the United States also constituted only a small fraction (less than 7 percent) of world registries during the period of its strong pressure for international pollution control (see table 8.3). The registries of Japan, the United Kingdom, and flag-of-convenience states like Liberia, Panama, and Greece all exceeded U.S. shares during the 1970s. Even if we consider Liberia-flagged tankers as effectively under U.S. control—since many are owned by American companies—the combined share has never exceeded one-third of all registries. U.S. control of registries may therefore explain some adoption of SBT, but it leaves much of this adoption unexplained.

Even in the area of oil imports, the United States has wielded less power than Japan since the mid-1960s. The United States has never controlled more than 20 percent of the market, and the U.S. share was at an all-time low during the 1970s when the United States was pressing hardest for SBT rules (see table 8.4). More tankers enter U.S. ports than the U.S. share of imports suggests, but port state inspections by the United States do not obviously explain the 98 percent compliance rates.[64]

Therefore, regardless of the metric used, the United States has exercised effective control of one-third of the world tanker market at best. The United States could have imposed its regulations on most tanker owners only by pressuring other states to adopt and enforce laws requiring SBT. Japan and most other states controlling construction, registry, and import markets never demonstrated an independent commitment to SBT. None would have adopted unilateral regulations requiring either COW or SBT, let alone both, without U.S. pressure. Yet there is no evidence of "arm twisting" on enforcement, and in any event enforcement would have been

63. M'Gonigle and Zacher, *Pollution, Politics, and International Law,* 121 and 137.

64. Even in general terms, most analysts argue that U.S. power was declining or staying even during the period when tanker owners were adopting SBT. See Joseph S. Nye, Jr., *Bound to Lead: The Changing Nature of American Power* (New York: Basic Books, 1990); and Paul M. Kennedy, *The Rise and Fall of the Great Powers: Economic Change and Military Conflict from 1500 to 2000* (New York: Random House, 1987).

more difficult in the absence of an international agreement. In fact, no states other than the United States have enacted legislation more stringent than the MARPOL requirements.

Although the United States used diplomatic pressures to reach agreement, it has never threatened to use or used sanctions against other states for failure to enforce MARPOL. In contrast, the United States has frequently threatened to impose trade sanctions for actions it views as "diminishing the effectiveness" of the International Convention for the Regulation of Whaling and the Convention on International Trade in Endangered Species.[65] Several countries have enforced MARPOL rigorously, including some that opposed the initial equipment regulations. However, the fact of their enforcement and the rules they have enforced do not reflect U.S. hegemonic pressures. The growing European environmental concern of the late 1970s led to the enforcement of MARPOL only after there was collaboration through the MOU, not because of U.S. pressures. Indeed many of these states voted against the U.S. SBT proposals in adopting both the 1973 and the 1978 agreements. The enforcement of the equipment standards and certification procedures these states were monitoring would also have been unlikely to result from a regional agreement. The MOU is specifically restricted to enforcing internationally agreed-upon standards. These states did not specifically want to enforce equipment standards; they wanted to enforce oil pollution standards, and the enforcement of MARPOL's equipment standards provided the easiest and most effective means available for doing so. Domestic political forces, not external hegemonic pressures, created the pressures for enforcement; regional collaboration created the conditions that fostered such enforcement; and MARPOL's equipment standards provided a more effective channel for enforcement than had its discharge standards.

Nongovernmental actors would also have responded quite differently if the United States had acted unilaterally. Classification societies and insurance companies would have been unlikely to establish universal

65. Dean M. Wilkinson, "The Use of Domestic Measures to Enforce International Whaling Agreements: A Critical Perspective," *Denver Journal of International Law and Policy* 17 (Winter 1989); and Gene S. Martin, Jr., and James W. Brennan, "Enforcing the International Convention for the Regulation of Whaling: The Pelly and Packwood-Magnuson Amendments," *Denver Journal of International Law and Policy* 17 (Winter 1989).

standards based on unilateral U.S. legislation. If the United States had passed and enforced unilateral SBT requirements, oil and oil transportation companies could easily have differentiated their fleets into U.S.-capable and non-U.S.-capable tankers, incurring SBT costs only on the number of tankers sufficient to service the U.S. market.[66] Passage of such unilateral legislation would also have faced stiff industry opposition, including potential boycotts and efforts to circumvent the rules.[67]

Even U.S. enforcement efforts reflect the influence of the equipment standards. Public goods theory predicts that actors will tend not to enforce rules that supply benefits to all other parties to a treaty. Axelrod and Keohane specifically argue that cooperation can be improved by increasing the "privatization" of benefits from enforcement.[68] Contrary to theory, however, the United States spends far more on enforcing equipment standards—a public good that improves the global ocean environment—than on enforcing discharge standards off its own coast, the benefits of which would be more private. It also seems unlikely that the United States would have detained tankers for breach of unilateral SBT requirements: although the United States had the practical ability to do so, without MARPOL such detentions would have constituted an infringement of flag states' sovereignty and international law. Indeed the United States began detaining ships only after MARPOL entered into force in 1983 even though the provision had been agreed to in 1973. Like the European nations, the United States enforced equipment standards because they were easier to enforce and more effective, despite the benefits they provided to other nations.

66. The unilateral U.S. requirements for SBT or COW on small tankers over fifteen years old allow us to estimate the impact of American unilateralism. The available data are not sufficiently detailed to ensure proper interpretation, but only 21% of such tankers have SBT or COW as compared to 94% of similar large tankers that face MARPOL requirements.

67. In 1990 the Royal Dutch/Shell Group announced plans to use small tanker companies to transport its crude oil to the United States as a means to protest, and protect itself from, laws that increased tanker operators' liability for damages due to oil spills; see John Evan Frook, "Both Arco and Unocal Say No Tanks to Shell Game," *Los Angeles Business Journal* 12 (25 June 1990), 3.

68. Robert Axelrod and Robert O. Keohane, "Achieving Cooperation under Anarchy: Strategies and Institutions," in Kenneth Oye, ed., *Cooperation under Anarchy* (Princeton, NJ: Princeton University Press, 1986).

In short, U.S. diplomatic pressures explain what rules nations agreed to in MARPOL 73/78, and U.S. power in oil markets explains why some tankers adopted SBT. However, hegemonic pressures alone would have resulted in SBT adoption rates far lower than the almost universal adoption actually observed, most likely well below 50 percent. MARPOL's equipment standards provided a focal point for monitoring, inspection, and detention efforts that facilitated both more frequent and more effective enforcement than were possible under the discharge standards.

Finally, increased public concern over oil pollution might explain increased tanker attention to discharges without MARPOL in two ways. First, if states had accepted progressively more stringent discharge standards in international agreements, then both the requirements and increased compliance after 1973 could indicate a general increase in international and social commitments to environmental values. However, the resistance to the U.S. proposals at the 1978 conference and the failure to tighten discharge standards document that no "sea change" had taken place in states' attitudes and commitment towards oil pollution control. Many states still opposed imposing significant costs on industries.

Second, growing public concern could have led oil companies to see increasing political benefits in reducing discharges. Large oil companies expend considerable effort to establish and maintain reputations as good corporate citizens. As environmental consciousness grew—in the United States in the late 1960s and in other countries through the 1970s—large oil companies increasingly had incentives to avoid intentional discharges that would soil company reputations along with birds and beaches. As early as 1963, the American Petroleum Institute noted that intentional discharges should concern tanker operators because they "may become a public nuisance and this is a reflection on the reputation of the shipping and petroleum industries."[69]

Increasing corporate concern should manifest itself in a gradual decrease in discharges through the 1970s and into the 1980s. Indeed, increased concern may have contributed to the decline in total discharges by large oil companies that was apparent in the LOT data (see figure 7.2). It seems likely that oil company tankers, not those of independents,

69. James E. Moss, *Character and Control of Sea Pollution by Oil* (Washington, DC: American Petroleum Institute, 1963), 46.

constituted more of the 36 percent of older tankers that installed SBT when they had the option of COW. It has been observed that "the oil majors, and to a lesser extent the other oil companies, are very much aware of their 'public image,' and are much more likely to retrofit their fleets as a matter of company policy, paying less attention to the individual costs involved"; in contrast, independents face much less forgiving economic circumstances and "would find it more of a sacrifice to retrofit their ships."[70] Even for oil majors, however, installing SBT became a necessary part of proving a company's "environmental credentials" only in the wake of MARPOL. Without such rules, oil companies and independents could have been expected to promote LOT and COW as cheaper and more effective pollution reduction technologies than SBT.

*   *   *

Comparing Discharge Limits and Equipment Standards: An Institutional Analysis

The all-but-perfect compliance with MARPOL's equipment requirements stands in marked contrast to the frequent violations of its discharge standards. Detailed empirical data as well as experts' opinions laid out in this and the previous chapter document the striking differences in compliance levels under these two compliance systems. Exclusive reliance on realist variables of power and interests fails to explain the wide variance in compliance levels between two rules in the same treaty. The two compliance systems provide a strong basis for comparison. Both compliance systems ultimately were intended to alter the same behavior, namely intentionally discharging oil. Both targeted the same nations and tankers. And both were legitimate and equally binding on treaty parties.[71]

70. Drewry, *The Impact of New Tanker Regulations*, 35.

71. Thomas M. Franck has recently suggested that variance in legitimacy may cause variance in compliance levels; see *The Power of Legitimacy among Nations* (New York: Oxford University Press, 1990). See also Philipp M. Hildebrand, "Towards a Theory of Compliance in International Environmental Politics," paper prepared for the Annual Convention of the International Studies Associa-

Oil pollution also represents a hard case. Theory predicts that treaty rules in this issue area would be highly unlikely to induce compliance. Collective action theory predicts that attempts to impose large pollution control costs on the powerful and concentrated oil transportation industry in order to provide diffuse, nonquantifiable benefits to the public at large would be unlikely to be adopted, let alone elicit meaningful behavioral change or compliance.[72] Besides, oil pollution is not a relatively easy-to-resolve coordination problem: continuing noncompliance and efforts to improve enforcement attest to the fact that the treaty rules were not self-enforcing, confirming that oil pollution control poses a difficult collaboration problem.[73]

Economic factors also make this a hard case: since the technology necessary for tanker operators to comply with the discharge standards—LOT—was cheaper and more cost effective than SBT, we should expect a higher level of compliance with the discharge standards, not the equipment standards. Finally, theories that the privatization of benefits increase the likelihood of enforcement would lead us to expect less enforcement of the equipment standards, with their global benefits, than of the discharge standards. Yet naval and aerial surveillance programs are few and small, while many states have added oil pollution equipment

---

tion, Atlanta, GA, 31 March–4 April 1992. Although it is possibly true in other cases, legitimacy fails to explain the variance observed here since it would predict higher levels of compliance with the discharge standards, which had wider support and a longer history.

72. Michael McGinnis and Elinor Ostrom, "Design Principles for Local and Global Commons," unpublished paper, Bloomington, IN, March 1992, 21. Mancur Olson's argument that small groups succeed in supplying public goods more often than do large groups relies on one or more members of the group receiving *personal* gain from providing the good, which is not the case with respect to oil transporters; see *The Logic of Collective Action: Public Goods and the Theory of Groups* (Cambridge, MA: Harvard University Press, 1965), 34.

73. Arthur A. Stein develops this concept very effectively in *Why Nations Cooperate: Circumstance and Choice in International Relations* (Ithaca, NY: Cornell University Press, 1990). See also Axelrod and Keohane, "Achieving Cooperation;" and Lisa L. Martin, *Explaining Multilateral Economic Sanctions: Coincidence, Coercion, and Coadjustment,* unpublished Ph.D. thesis, Harvard University, Cambridge, MA, 1989.

control to port state inspection procedures, and regional port state control regimes are increasing in number.

Despite the fact that oil pollution control is a hard case, and despite the inability of parsimonious arguments based on economic interests or international political power to explain the high level of compliance with the equipment standards, claiming that the compliance system induced the observed compliance demands that we identify how it did so.

The compliance systems for the equipment and discharge standards differed most strikingly in the fundamental model underlying each of the strategies. Enforcement of the equipment standards relied on a coerced compliance strategy that was intended to monitor behavior before violations were committed to prevent them. Enforcement of the discharge standards was deterrence-oriented, intended to detect violations after they occurred and to prosecute and sanction violators to deter them from committing violations in the future.[74] This basic difference in orientations itself made the task facing the compliance system for the equipment standards more manageable than that facing the compliance system for the discharge standards. A preventive orientation was more congruent with the equipment standards, but not inherent in them; a coerced compliance strategy involving independent observers assigned to ride on all oil tankers and monitor discharges had been proposed and could have been developed.[75] The fundamental strategy choice had important consequences for the level of compliance likely: inhibiting an actor's ability to violate proves far more effective than increasing the disincentives for violating.

The equipment standards also elicited significantly higher levels of compliance because the compliance system ensured that at least some actors had the political and economic incentives, practical ability, and legal authority to perform the tasks required for compliance. The new

74. Notably neither strategy involved in incentive-based or "voluntary" strategy, like the funding of compliance under the Montreal Protocol and the Framework Convention on Climate Change. For a development of the distinction between these three strategies, see Reiss, "Consequences of Compliance"; and Keith Hawkins, *Environment and Enforcement: Regulation and the Social Definition of Pollution* (Oxford, England: Clarendon Press, 1984).

75. Inspectors were estimated to cost $100,000 per ship per year, and they would prove more economically efficient since they left technology choices "to those who could be expected to be most familiar with relative costs" (Cummins et al., "Oil Tanker Pollution Control," 171).

primary rules laid the basis for a more effective compliance information system that refinements of the discharge standards could not have achieved. There were actors that could and would monitor tankers during construction and retrofits as well as after tankers began trading. There were also actors that were able and willing to impose significant costs on those that sought to violate the treaty rules or already had violated them. For all these tasks, the rules succeeded in placing actors in an incentive-ability-authority triangle.

The equipment standards greatly simplified the task of identifying violations both before and after they were committed. By focusing on the tanker builder–tanker buyer transaction, the rules drastically reduced the number of acts, actors, and locations that needed to be monitored. Although there are some 7,000 tankers, there are far fewer tanker owners, and only a few major shipyards and classification societies. Therefore, initial surveys and periodic inspections could be readily undertaken and have a reasonable likelihood of being comprehensive.

Evidence of equipment violations and links to violators did not dissipate over time as they did with discharge violations. A tanker captain discharging illegally knew that, if the discharge was not detected immediately, there was little chance of its being detected at all. In contrast, tanker owners inclined to violate the equipment standards knew that a violation would be readily visible throughout the tanker's life. The "passive-voice" violations that plague enforcement of the discharge standards—discharges that cannot be linked to specific tankers—were eliminated under SBT and COW. The equipment standards reassured authorities that they could easily detect violations and identify violators; they needed to monitor tankers only during business hours in port rather than over wide expanses of sea at all hours. Better yet, these simpler and cheaper programs provided dramatically greater assurance that most violations were being detected. The high visibility of these standards also reassured tanker owners that were inclined to comply that other tanker owners would not get away with equipment violations, gaining a competitive advantage.

In contrast to the discharge provisions, the equipment standards did not rely on prodding reluctant states into making greater enforcement efforts, but gave those developed states with domestic political incentives to monitor compliance the practical ability to do so. Although verifying an oil record book was difficult, verifying an IOPP certificate was easy once port states had the legal authority to conduct thorough inspections. Many nations that now had the practical and legal ability to verify IOPP certificates against actual equipment also had to do so. Although the incentives to undertake enforcement have generally come from domestic publics and nongovernmental organizations, these pressures have resulted in more effective detection of equipment standard violations than the same pressures and resources dedicated to enforcing the discharge standards.

Additionally, the equipment standards made monitoring easier and more likely by relying on existing infrastructures. Long before MARPOL established pollution equipment requirements, classification societies and insurance companies had been monitoring compliance with many nonenvironmental international standards. Piggybacking on existing infrastructures meant that these nongovernmental actors could effectively monitor compliance with the new rules with only marginal changes to incorporate pollution equipment inspections into their standard operating procedures. In contrast, discharge detection had required completely new surveillance programs. The compliance system for the equipment standards also piggybacked on existing port inspections for adherence to safety and customs rules. Like classification societies, port authorities traditionally inspected tankers for compliance with other international rules, so pollution prevention was simply added to the standard inspection routine of inspectors. Indeed pollution prevention remains only one of many elements in the port state control efforts conducted by the European MOU states, the United States, Canada, Japan, and other states.

MARPOL's equipment standards also made preventing and sanctioning violations easier. Preventing equipment violations proved far easier than preventing discharge violations because it involved a transaction between actors with different incentives. Tanker buyers faced a few infrequent decisions requiring cooperation from other actors and involving major economic consequences. A shipbuilder could easily prevent a buyer from acquiring an illegally equipped tanker, and classification societies and insurance companies could easily prevent such a tanker from operating on the international market. In contrast, tanker captains faced many frequent decisions about whether to violate discharge standards, and no other actor had the ability to prevent such violations. For port state authorities, the administrative sanctions of detention or barring a ship from port were also far easier to effectively implement since these santions skirted both their own domestic legal system and that of any flag state. They required less time, effort, and coordination than the legal process of investigating, prosecuting, and penalizing a discharge violations.

Not only were such measures easier, but they provided the foundation for a noncompliance response system involving far more potent sanctions than those available for discharge violations. New tankers "cannot get insurance without [IOPP] certification, and can't get certification without compliance."[76] Thus compliance became a necessary prerequisite to doing business, an exceptionally potent sanction in itself. Even if this sanction failed, the ability of governments to detain a noncompliant tanker or bar it from port imposed opportunity costs that could in a single day exceed the maximum fines allowed in many nations. Once again, compliance became a prerequisite to trading. Finally, "anyone who buys a ship wants to have a valuable asset, and if you build a ship that can't trade anywhere then you diminish the asset's value."[77] Tankers are frequently sold, and "conformity to international standards is an important bonus when it comes to selling a ship."[78]

Besides simplifying the tasks involved in the compliance information and noncompliance response systems, MARPOL gave authority to actors that were likely to use it. Recognizing that many flag states lacked either the ability or the incentives to survey and inspect tankers, the drafters of MARPOL allowed governments to delegate such responsibilities to classification societies. From a flag state's perspective this was attractive because it eliminated a task for which it would otherwise be financially responsible. From the enforcement perspective this was beneficial because it empowered "front-line" inspectors who had both a greater ability and

76. Interview, John Foxwell, Shell International Marine, London, England, 27 June 1991.

77. Interview, Daniel Sheehan, U.S. Coast Guard, Washington, DC, 9 April 1992.

78. Bergmeijer, "The International Convention for the Prevention of Pollution from Ships," 12; and Interview, Ton IJlstra, Interdepartmental Coordinating Committee for North Sea Affairs, the Hague, the Netherlands, 13 June 1991.

a greater likelihood to undertake consistent and thorough inspections and surveys while removing the process from the vagaries of governmental budget fluctuations. Classification societies had incentives to monitor treaty compliance in order to protect their reputations in competing for business and to avoid legal battles with insurance companies. Government monitoring of classification societies' issuance of IOPP certificates reinforced the societies' incentives to monitor carefully and accurately.

Government monitoring increased relative to the monitoring of compliance with the discharge standards because MARPOL provided developed port states with the legal authority to conduct the necessary inspections. In contrast, developing states, especially oil-loading states, had proven consistently unwilling to conduct the inspections needed to verify compliance with the total discharge standards. Such monitoring was also made more attractive to the developed states since they now had the right to follow through on detected violations; the right to detain a ship eliminated the need to turn over evidence of a violation to a flag state that might well not prosecute. Detention was not so excessive as to deter governments from imposing it. Whereas states had proven unlikely to impose high penalties for discharge violations, administrative detention seemed appropriate for equipment violations, but imposed on tanker operators opportunity costs sufficient to deter others from committing violations. As with monitoring, the frequency of the imposition of sanctions was increased not by convincing reluctant governments to enforce, but by removing the legal barriers inhibiting the effectiveness of enforcement from those states willing to enforce.

A final word is in order on this comparison of discharge and equipment requirements. The failure of discharge standards and the successful forcing of SBT on a reluctant industry exhibit "the classic issues in environmental control policies generally. On the one hand, mandatory design requirements may miss least-cost solutions and inhibit innovation. On the other hand, they offer greater certainty of compliance and reduce enforcement costs."[79] Although discharge standards alone allowed for the flexibility essential to economic efficiency, they failed to provide a means of effecting behavioral change. SBT and COW requirements ensured such change and even made discharge standards redundant for the

79. Pearson, *International Marine Environmental Policy*, 97.

tankers affected, but risked incurring costs that exceeded the social benefits of the resulting reduction in pollution.[80] Other approaches, like liability schemes and placing inspectors on board every tanker, could conceivably achieve the same results at significantly lower cost, would encourage the innovation of more effective pollution abatement equipment and procedures, and would achieve greater economic efficiency.[81] However, it has been precisely those equipment requirements, with their better compliance system, that have been the main tools that have made MARPOL effective.[82] Equipment requirements have clearly induced greater compliance than have the discharge requirements; comparisons with respect to the efficiency, cost effectiveness, and overall social benefits of the two approaches might result in a different ranking. The task of this research, however, has been to clarify whether treaty rules can influence behavior.

Conclusions

Treaty rules can induce compliance. The SBT case provides convincing evidence that treaty rules can affect behavior. MARPOL's equipment requirements and the associated compliance system completely altered the way oil transport companies operate. Compliance is almost perfect. Despite strong incentives not to install SBT, tanker owners have done so as required by MARPOL. When they are not required to do so or when they have a choice between SBT and COW, they have overwhelmingly chosen not to install SBT. Even when economic factors, unilateral U.S. requirements and pressures, and public environmental concern are accounted for, the all but universal adoption of SBT remains explicable only as behavior that would not have occurred absent the MARPOL

80. Pearson, *International Marine Environmental Policy,* 96.

81. Such "emission-on-charge" systems have been criticized, however, even at the domestic level because enforcement requires "regular interactions between the government and every single polluter rather than reliance primarily on the credible threat of enforcement to induce 'voluntary' compliance"; see Steven Kelman, *What Price Incentives?: Economists and the Environment* (Boston, MA: Auburn House Publishing Co., 1981), vii.

82. Robert Blumberg, U.S. State Department, confirmed that "what works are mandatory equipment requirements"; Interview, Washington, D.C., 20 May 1991.

requirements. In contrast, OILPOL and MARPOL's discharge standards had far less impact. Noncompliance remains common.

The marked contrast between the responses to the two sets of standards by the same governments and companies with the same interests during the same time period can be explained only by examining the compliance systems in which the different standards are embedded. Although the compliance system for the equipment standards had the ability to reliably detect and sanction violations, the very low levels of noncompliance with these standards strongly suggests that the compliance system has succeeded by preventing violations from being committed in the first place rather than by deterring violators from committing them in the future. This coerced compliance strategy has elicited far higher compliance levels than the deterrence-based strategy used to enforce the discharge standards. New tankers have been built to MARPOL standards from the beginning, not forced to retrofit later. The discharge compliance system created primary rules that either had little practical ability of being enforced or of being monitored or left the practical ability and legal authority to enforce them and monitor compliance with governments and nongovernmental actors that had no incentives to do so. Not surprisingly, tanker captains often remained undeterred from making illegal discharges. The equipment standard, in contrast, succeeded by ensuring that the actors that would be responsible for preventing, and if necessary detecting and sanctioning violations, had both the incentives and the authority to do so. It increased the practical ability and legal authority of those governments and private actors that already had incentives to monitor and enforce, while simultaneously reducing the efforts needed to do so. These improvements, in turn, altered the compliance incentives facing tanker owners. Through these design features, the compliance system supporting MARPOL's equipment standards has produced an extraordinary change in how oil tankers are constructed, a change that would not have been made in the absence of international environmental law.

IV

Conclusions

What major conclusions can we draw from the array of different treaty rules that have been adopted to change government and industry behavior to reduce intentional oil pollution? What findings and propositions hold true across the several rule changes examined here? How should policymakers frame future international environmental agreements so they induce greater compliance? This final section answers such questions, delineating both the book's major findings and a series of policy proposals for practitioners. It evaluates how accurately the results of a study of intentional oil pollution can be generalized to other environmental arenas, while identifying the limits to such generalizations. It then evaluates whether the efforts to regulate intentional oil pollution have achieved effective environmental protection as well as compliance. The chapter and the book conclude by identifying areas in which future research is needed.

9

Conclusions: Uniting Theory and Practice

Nations will continue to negotiate treaties to address international environmental problems for the foreseeable future. Whether those treaties improve the management of those problems will depend on how negotiators frame treaty proscriptions and prescriptions and on the types of compliance systems they establish. This book has shown that treaties can lead powerful governmental and corporate actors to adopt new behaviors that they initially opposed. Experience with several efforts to improve the compliance systems of the International Convention for the Prevention of Pollution of the Sea by Oil (OILPOL) and the International Convention for the Prevention of Pollution from Ships (MARPOL) refute claims that "considerations of power rather than of law determine compliance."[1] After accounting for the incentives and power that states have to violate, we can attribute the remaining changes in behavior and variance in compliance levels to the success of treaty provisions in accomplishing three tasks: creating "opportunistic" primary rule systems that impose requirements on those actors most likely to fulfill them, creating compliance information systems that give information providers a "return on their investment," and creating noncompliance response

1. Hans Joachim Morgenthau, *Politics Among Nations: The Struggle for Power and Peace* (New York: Alfred A. Knopf, 1978), 299. *International Convention for the Prevention of Pollution of the Sea by Oil*, 12 May 1954, 12 U.S.T. 2989, T.I.A.S. no. 4900, 327 U.N.T.S. 3, reprinted in 1 I.P.E. 332, hereinafter cited as *OILPOL 54*. *International Convention for the Prevention of Pollution from Ships*, 2 November 1973, reprinted in 12 I.L.M. 1319 (1973), 2 I.P.E. 552; and *Protocol of 1978 Relating to the International Convention for the Prevention of Pollution from Ships*, 17 February 1978, reprinted in 17 I.L.M. 1546 (1978), 19 I.P.E. 9451, hereinafter together cited as *MARPOL 73/78*.

systems that remove international legal barriers from those actors with incentives to respond to noncompliance.

Rather than ignoring the power and interests of relevant actors, negotiators have elicited high levels of compliance when they have taken advantage of the fact that power and interests vary from treaty provision to treaty provision. Overall levels of environmental concern and the resources governments are willing to dedicate to protecting the environment will set broad limits on compliance levels. Policy leverage comes from the recognition that the resources dedicated to implementing a treaty and the effectiveness of those resources are not fixed for an issue area, but depend on specific characteristics of how the treaty's provisions are framed. The choices policymakers make regarding how to define and constrain compliance, monitoring, and enforcement have significant implications regarding how much compliance the treaty elicits.

Although oil pollution is a sufficiently important environmental problem to warrant study in its own right, the real value of this book's findings will stem from applying the lessons learned here to other international environmental problems in the hope that some failures may be avoided and some successes repeated. This chapter's first section develops several theoretical themes that synthesize the conclusions of the empirical work in chapters 4 through 8. It frames these major findings in terms that address the larger theoretical debate outlined in chapter 2 while allowing their extension to the broad range of environmental problems of which intentional oil pollution regulation is just one example. The chapter's second section delineates several detailed policy prescriptions that stem from the experience with intentional oil pollution. Each of these recommendations (highlighted by italicized text) follow from the evidence laid out in preceding chapters, but adapt the lessons learned from this evidence so they can be applied in other environmental arenas. Both these tasks assume that lessons from oil pollution are not so unique as to be irrelevant to other issue areas. The chapter's third section addresses this concern directly, identifying the ways in which the oil pollution experience is similar to the experience of other issues, supporting extension of the book's findings. I also note the contextual factors and unique elements of this experience that conditioned the success of treaty rules and limit the generalizability of the findings. The chapter's final section discusses the relationship of compliance to effectiveness and delineates the questions for further research that have been provoked by this book's study of the impact of treaties on behavior.

Theoretical Themes

The Importance of a Treaty's Rules

This book started by asking, "Can treaty rules induce behavioral change?" The evidence presented unequivocally demonstrates that governments and private corporations have undertaken a variety of actions involving compliance, monitoring, and enforcement that they would not have taken in the absence of relevant treaty provisions. Compliance is not merely a reflection of power. There is a wide variance in compliance levels even across two rules targeting the same underlying behavior within a single issue area. Factors exogenous to the treaty leave much of this variance unexplained. The remaining variance fits well with causal models showing how treaties and compliance systems have induced governments and industry to adopt otherwise undesirable behavior.

The cynical view introduced in chapter 1 and developed in chapter 2 that rules have no independent causal relationship to compliance implies that compliance with treaty provisions cannot be increased by rule changes that are made independent of changes in the underlying determinants of compliance, namely power and interests. This study provides information on three clear cases in which a wide array of actors complied with rules regulating one behavior while at the same time failing to comply with other rules regulating essentially similar behavior. The first and most striking example involved the contrast between compliance with discharge limits and compliance with requirements for tankers to install segregated ballast tanks (SBT). Essentially every tanker required to install SBT did so on the schedule delineated in MARPOL during the same period of time that detected tanker discharges remained commonplace, showing no identifiable decline. The equipment rules led tanker owners to install SBT despite significant costs, the absence of economic benefits, and decreasing oil prices that were increasing pressures to cut costs. These very factors explain why the majority of tankers exempt from the equipment requirement have not installed SBT and affirm the conclusion that the installations represented treaty-induced compliance. Although many tankers were registered in states that opposed the adoption of the SBT requirements and had strong incentives not to comply, all those required to comply did so.

Governments often did not change their behavior, but industry did. It did so not as an agent of any state, and in most cases not even in response to enforcement threats by flag states. Rather universal SBT adoption reflected a direct response to international rules based on a "coerced compliance" model of regulation that was intended primarily to prevent violations rather than deterring potential violators from committing violations in the future. The rules facilitated initial surveys and inspections for such equipment by nongovernmental classification societies that made it quite difficult for a tanker to receive classification and insurance papers or to trade internationally without the proper equipment on board. The treaty reinforced this system by establishing more effective in-port monitoring and enforcement, the former of which has increased significantly. Although oil price fluctuations and hegemonic exercises of power by the United States help explain the timing of SBT rule adoption, and although unilateral U.S. enforcement efforts would have led some tankers to install SBT, most SBT adoption must be viewed as treaty-induced compliance. International requirements alone explain the installation of SBT on most tankers. The continuing failure of operational standards to elicit high levels of compliance bolsters the contention that it was the superior compliance features of the SBT requirements rather than other factors that induced the behavior change.

A comparison of the responses by European states to the enforcement reporting systems of the International Maritime Organization (IMO) and the Memorandum of Understanding (MOU) on Port State Control provided a second case in which unambiguous evidence demonstrated that compliance systems matter. The same states during the same time period provided only half as many reports on enforcement of MARPOL requirements to the IMO Secretariat as they did to the MOU Secretariat. The only possible explanation for such a wide discrepancy has to do with the different reporting systems established by the two organizations. The MOU system proved far more successful because it succeeded in incorporating its reporting requirement into the standard operating procedures of the enforcing bureaucracies and because it processed data in ways that reinforced the reporting authorities' interests in effectively using their enforcement resources. The IMO system failed to give those with the ability and authority to inspect and report any incentives to do so, while the MOU system created such incentives by providing an information tool that made inspection programs easier and more effective to operate.

Prior to the entry into force of the MARPOL treaty in 1983, no government had ever detained a foreign-flagged tanker for violating oil pollution laws. In contrast, since 1983 several states have detained tankers for MARPOL violations. The different responses to MARPOL's requirement that states detain any tanker posing a threat to the marine environment highlight the conditions under which the new requirement influenced behavior. Most states have not detained a single ship even since 1983, demonstrating that the rule has not transformed reluctant and unconcerned states into rigorous enforcers. Rather the new rule removed the legal obstacles that prevented those with preexisting incentives to enforce from doing so rigorously. Prior to MARPOL's entry into force, these states felt constrained by existing international law that banned such detentions. The treaty change did not alter the practical capacity of these states or their political incentives to undertake such drastic measures, but it did change their legal standing (and hence, presumably, their international political costs), making the use of such measures more likely and thereby increasing the deterrent value of enforcement efforts.

In these three cases, clear causal links unambiguously demonstrate that treaty rules independently influenced behavior, with other plausible factors controlled for or absent. These cases only place a lower limit on the treaty's impact on behavior, however. In several other cases, behavioral changes appear to have resulted from treaty requirements and procedures, but the complexity of the causal links prohibits us from isolating that fraction of the change that was due to treaty requirements from the fraction that was due to other factors. The following discussion suggests other cases in which treaty influence seems plausible, but the evidence is weaker.

How a country's environmental concern expresses itself in legislative and enforcement activity is also a function of international rules. After MARPOL was signed, a government could no longer claim that discharge standards constituted a strong environmental policy. Although domestic industries oppose standards more stringent than a treaty's, environmental lobbies color legislation less stringent than international commitments as environmentally malevolent.[2] Treaties frame the debate over responses to environmental problems in terms different than would be used otherwise. States like Japan and France, which opposed equipment standards in 1973 and 1978, implemented and enforced those standards without pressure from the United States, actions that were unlikely in the absence of international agreement. Cooperation with the MOU did not require hegemonic imposition by the United States: many equally powerful states overcame free-rider concerns by establishing an information system that benefited member nations. U.S. pressures for enforcement led to new enforcement efforts by these governments only in the indirect sense that U.S. pressures in 1973 and 1978 led to MARPOL's equipment standards and, in agreeing to enforce IMO agreements, these governments did not exclude MARPOL's requirements.

Most analysts contend that the total amount of oil entering the ocean has decreased since rules regulating oil discharges were first agreed to in 1954. Although accurate figures remain elusive and illegal discharges remain common, much evidence suggests that many tanker operators and oil transportation companies act more conscientiously about discharging oil at sea than one would expect without international regulation. Numerous factors undoubtedly have contributed to these changes. Rising oil prices and cheaper techniques for generating less waste oil have played a role. Growing environmental consciousness and norms against polluting undoubtedly have led some tanker captains and companies to decide to discharge waste oil in reception facilities or use load on top (LOT) technology rather than dumping waste oil at sea. Although the OILPOL and MARPOL rules were certainly a product of environmental concern, they subsequently have reinforced it, contributing to general social pressures on captains not to discharge oil at sea. Thus treaties may contribute to a process of education and altering the values pursued by countries, companies, and individuals.

Development of LOT demonstrates another pathway of influence. In developing LOT, companies developed discharge-reducing processes in

2. Only the U.S. imposed tighter equipment deadlines than were called for by MARPOL; see James Cowley, "The International Maritime Organisation and National Administrations," *Transactions of the Institute of Marine Engineers* 101 (1989), 134.

direct response to treaty pressures. Although the procedure violated existing international rules, its adoption was prompted by efforts to tighten those requirements and would not have occurred otherwise. The absence of oil price increases until well after LOT was developed confirms that economic considerations did not prompt the new research. Oil companies devised a new means of achieving their economic objective while also achieving environmental objectives. The emissions trading and tax proposals that are frequently promoted to address climate change problems are aimed at providing incentives for precisely this type of learning and production process development.

Reception facility requirements provide a final example. Compliance has been spotty, with facilities remaining underprovided in much of the world. Nevertheless, there are considerably more reception facilities today than when OILPOL was first signed in 1954. Although precise causes prove difficult to isolate, IMO rules along with ongoing IMO attention to the need for and means of providing reception facilities at meetings, in seminars, and through surveys undoubtedly underlie the provision of many reception facilities. Without such efforts, although domestic political factors and economics would have led many governments and companies to install some facilities, fewer facilities would currently be available.

The Frequency of Significant Noncompliance

Demonstrating that treaty rules can influence behavior by no means suggests that they always do. In several cases, treaty rules had little or no impact on behavior. Despite the common assertion that most countries comply most of the time, noncompliance with several provisions of treaties regulating intentional oil pollution does occur, and indeed is quite common.[3] Although "most of the time" is difficult to define, this study provides clear evidence that noncompliance is often more than anecdotal.[4] Rules limiting both rates and total amounts of discharges, rules

---

3. See note 18 in chapter 1.

4. Abram Chayes and Antonia Chayes argue that "although there are some obvious exceptions where states have signed treaties without a serious intention to comply, ordinarily the decision is made in good faith"; see "Compliance without Enforcement: State Behavior under Regulatory Treaties," *Negotiation Journal* 7 (July 1991), 311.

requiring stringent penalties, obligations that states monitor and enforce a treaty, and requirements for reporting have all been frequently flouted. Indeed states make commitments that they subsequently fail to fulfill. Treaty membership cannot be safely assumed to imply compliance.

These examples do not undercut the claim made here, however. Many factors explain why treaty rules may fail to achieve the nominal objective of behavioral change. In some cases, as I have argued with respect to requirements to increase penalties, rules have been "designed to fail," agreed to precisely because they do not require any identifiable change in current patterns of behavior. In other cases, such as those involving requirements to install reception facilities by certain deadlines or rights to inspect tankers for violations of total discharge limits, failure to alter behavior was due to a failure to address the major causes of noncompliance. To refute the contention that treaty rules do not influence behavior required only that one find clear, empirically supported cases in which treaty rules did influence behavior, independent of other factors.

Although noncompliance is common, developed states appear to comply more often than developing states. Whether an absence of environmental concern or an absence of administrative capacity and resources causes this noncompliance remains unclear from the study done here. Indeed probably both factors contribute to the problem. The evidence presented here supports the view that the weak take advantage of the propensity of the strong to provide public goods in the international realm rather than the view that the strong force the weak to comply while reserving their own right to ignore treaty "dictates whenever it suits their purposes."[5]

A related finding is that government compliance appears to be more difficult to elicit than industry compliance. Although tanker operators have complied with quite expensive equipment requirements, compliance with the less costly reception facility requirements—a burden arguably placed on governments—has been far less frequent. Where there has been

---

5. Compare Mancur Olson, *The Logic of Collective Action: Public Goods and the Theory of Groups* (Cambridge, MA: Harvard University Press, 1965), 35; and Oran Young, "The Effectiveness of International Institutions: Hard Cases and Critical Variables," in James N. Rosenau and Otto Czempiel, eds., *Governance without Government: Change and Order in World Politics* (New York: Cambridge University Press, 1991), 187.

compliance, it has often been due to government regulations that require industry to provide these facilities. Governments have enforced discharge provisions so that tanker operators have had to absorb the costs of the equipment. This suggests that governments comply when they can shift the costs and burdens of compliance to industry.

Nonreporting: Common but Not an Indication of Noncompliance

Related to these findings, the evidence of reporting on enforcement and reception facilities clearly illustrates that nonreporting is common and noncompliance cannot be safely inferred from nonreporting. Self-reporting has increasingly been considered a key first step to inducing compliance with environmental treaties.[6] Yet even states that comply with substantive provisions may fail to report that fact. Although all the MOU states had extensive enforcement campaigns, half of them failed to report this activity to IMO.[7] The fact that all these states are industrialized, developed states undercuts the frequent argument that administrative incapacity is the major reason for nonreporting. On the other hand, surprisingly states sometimes provide evidence of their own failure to meet the substantive requirements of an agreement. Again the MOU experience is an example of a case in which all member countries have reported on their inspections, but only a very few have met the 25 percent inspection rate criterion stated in the MOU agreement. Similarly, several states have reported on ports that do not have reception facilities.

Sources of Treaty Influence

Comparing these cases highlights the wide variance in compliance levels even within a single issue area. Across environmental treaties, even

6. See, for example, Abram Chayes and Eugene B. Skolnikoff, "A Prompt Start: Implementing the Framework Convention on Climate Change," a report from the Bellagio Conference on Institutional Aspects of International Cooperation on Climate Change (Cambridge, MA: 28–30 January 1992); and United States General Accounting Office, *International Environment: International Agreements Are Not Well Monitored* (GAO/RCED-92-43) (Washington, DC: GPO, 1992).

7. Other cases in which enforcement has gone unreported to IMO are delineated in chapter 4.

greater variance in compliance levels can be found.[8] More important, the cases studied show that the level of compliance is causally contingent upon the type of rules and procedures adopted. Treaty provisions have altered behavior when they have placed at least some actors within what we might call an incentive-ability-authority triangle.[9] This triangle is conceptualized in figure 9.1.

In each of the three successes, treaty provisions provided the missing element of this triangle. Although tanker owners had always had the ability and authority to install SBT on their tankers, the new rules created the incentive for them to do so by piggybacking inspections for such equipment on existing classification society inspection procedures. This made the ability to trade internationally contingent on compliance with MARPOL equipment requirements, a very strong incentive indeed. Equally important, the tanker procurement process provided the context for a coerced compliance system in which the ability to violate was actually reduced. The fourteen European governments that were members of MARPOL and the MOU had equivalent ability and authority to conduct and report on enforcement activities under both agreements. The MOU Secretariat has received reports from all fourteen states every year, while IMO has received only half as many, because the MOU made reporting worthwhile and easy—worthwhile because the computerized system established under the MOU enabled states to deploy inspectors more effectively, and easy because the Secretariat of the MOU maintained the information system. Rules allowing detention changed the enforcement tactics of some states by giving those states that already had the incentives to sanction vigorously the legal authority to do so while increasing their ability to do so by replacing legal with administrative sanctions.

In contrast, treaty provisions failed to improve compliance when they ignored this strategic triangle. Discharges have remained common because the compliance system failed to deter potential violators from committing such actions while leaving their ability to discharge unimpaired. The total discharge limits were never monitored because initially

8. See Simon Lyster, *International Wildlife Law: An Analysis of International Treaties Concerned with the Conservation of Wildlife* (Cambridge, England: Grotius Publications, 1985).

9. See figure 9.1.

Appropriate incentives

Legal authority   Practical ability

Figure 9.1
The strategic triangle of compliance

only flag states, shipping companies, and oil companies had the authority to inspect tankers, and none had incentives to do so. Providing inspection rights to port states failed to remedy the problem because the oil-loading port states that had the ability to inspect tanks after a ballast voyage also had no incentives to do so. Efforts to increase penalty levels failed because states already had the ability and the authority to raise penalties, and the rules did not change the incentives. Reception facilities remain unprovided in many oil-loading states because these states certainly lack the incentives, and in some cases may lack the finances, to provide them.

Whether one is dealing with issues of compliance, monitoring, or enforcement, if the rules do not place actors within an incentive-ability-authority triangle, behavior will not change. The failure to alter behavior will, in turn, result in compliance levels that will be lower than otherwise would be possible. The drafters of treaties should not aim or expect to achieve perfect compliance. Nonetheless, they can fail to elicit potential compliance if analytic failures leave important components of the compliance system unaddressed.

The Two-Level Nature of Compliance with Environmental Treaties
The experience with the oil pollution treaties highlights the fact that achieving compliance with environmental treaties requires altering the behavior of nonstate actors, whether corporations or individuals, as well as governments. In the environmental realm, the usual focus of the international relations scholar on state—read government—behavior proves misplaced. Understanding why states comply with environmental treaties requires us to open up the "black box" of the state to examine

the roles of governments, industry, nongovernmental organizations (NGOs), publics, and individuals.

Many environmental problems have their sources in externalities of industrial and individual behavior. The behavior the treaty is intended to change is at the subnational level. For that reason, this book illustrates the role that governmental enforcement of international rules against private, nonstate actors plays in inducing such behavioral change. Governmental enforcement against private actors, or at least its threat, has proved a critical prerequisite to compliance. The absence in international law of centralized enforcement, often cited as the reason for violations whenever compliance proves inconvenient, does not imply the absence of enforcement. As the right to detain tankers demonstrates, international law can create conditions of decentralized enforcement by a few states that create pressures for compliance that extend far beyond the borders of those states. Eliciting compliance may require that an environmentally concerned government succeed in inducing laggard governments to pass and enforce treaty-implementing legislation that, in turn, succeeds in inducing subnational actors to change their behavior. The ongoing failure to induce reluctant governments to adopt more rigorous enforcement programs highlights the additional obstacles that the two-level nature of enforcement introduces.[10] Nevertheless, enforcement may be increased by removing the international legal constraints that prevent leader governments from directly influencing these subnational actors.

Not only are nonstate actors the ultimate targets of much international environmental regulation, but the oil pollution experience also provides examples of the important role NGOs play in the compliance process, especially the compliance information system. As was highlighted in chapter 8, the success of the equipment requirements has depended heavily on the involvement of classification societies, insurance companies, and tanker builders in the tanker procurement process. Their involvement has provided the essential foundation for a coerced compliance strategy that has prevented violations from being committed while simultaneously dramatically reducing the monitoring demands placed on governments. Nongovernmental actors also have played im-

10. Robert D. Putnam, "Diplomacy and Domestic Politics: The Logic of Two-level Games," *International Organization* 42 (Summer 1988).

portant roles in collecting, analyzing, and disseminating compliance and enforcement data reported by governments to IMO. Indeed many of this study's compliance data were collected and analyzed not by governments or IMO, but by nongovernmental actors. Data on compliance with SBT and COW standards came from independent industry research groups. Tanker trade organizations conducted studies of inadequate reception facilities. Secretariats may lack either the resources or the incentives to collect, analyze, and distribute compliance and enforcement data. NGOs and industry pick up this slack. The important contributions that nongovernmental actors can make to compliance information systems has been recognized by the U.N. Human Rights Commission and by the recently established Commission on Sustainable Development.

The oil pollution experience also suggests the difficulty such actors have in effectively responding to noncompliance and non-enforcement. They often lack the resources and the legal authority to use those resources in ways that pose potent and credible threats to governments. Although national governments are not the only important actors on the international scene, they continue to dominate the legitimate use of force and coercion. Even in issue areas in which NGOs have taken action, such as whaling and debt-for-nature swaps, their actions have tended to address only a small fraction of the problem. Therefore they need to take advantage of education and publicity as means of bringing the more powerful force of public opinion to bear on governments to change their behavior.

The Rare Use of Reciprocity to Elicit Compliance

Reciprocity has frequently been touted as the major means used by governments to induce other states to comply with international commitments.[11] If the term *reciprocity* is used to mean a strategy of using compliance to reward other compliers and violation to sanction other violators, then the cases examined here provide little support for this proposition. In the area of oil pollution, states have not chosen to make their own compliance with substantive treaty provisions contingent on

11. See, for example, Robert O. Keohane, "Reciprocity in International Relations," *International Organization* 40 (Winter 1986); and Robert Axelrod, *The Evolution of Cooperation* (New York: Basic Books, 1984).

the compliance of other states. Other factors determine whether a state complies with and enforces an agreement, and these factors are not linked to the compliance or enforcement of other states. France proposed that discharge limits be waived for tankers traveling to ports not complying with reception facility requirements, but the proposal received no significant support. In dealing with issues like arms control or trade, retaliatory noncompliance may provide a means of recouping losses imposed by another states' noncompliance. In the area of environmental affairs, environmental states appear unwilling to reduce their own compliance or enforcement efforts to coerce others. Rather these states resort to linking target issues with other issues rather than to specific reciprocity. In standard prisoners' dilemma language, environmental leader states do not face binary choices between compliance and defection, but rather choose compliance independent of others' actions and then choose whether to impose sanctions or provide positive incentives to induce compliance by others.

Difficulties in Disseminating Information about Compliance

A final finding is that oil pollution secretariats prove surprisingly unwilling to disseminate information that would allow interested actors to respond to noncompliance. IMO has only infrequently conducted analyses of the few enforcement reports it has received. Little effort is made to identify nonreporting or noncompliance. Reception facility data have been largely analyzed and published without identifying the ports required to have reception facilities. The MOU Secretariat's annual reports do not provide the data needed to identify the countries that are failing to meet the 25 percent inspection requirement. The MOU Secretariat has also consciously decided not to release a blacklist of tankers found violating MARPOL regulations to the private oil companies and chartering organizations whose failure to do business with such tankers could provide strong incentives for such tankers to comply.

The failure to analyze existing data arises in part from inadequate resources to cover myriad organizational priorities. The failure to disseminate the information as a means of prompting negative media and NGO publicity, public pressure, and sanctions by private companies is also due to diplomatic deference to other governments and the fear of legal problems due to releasing inaccurate information about specific corporations. Unfortunately governments' willingness to provide reports may prove inversely related to the likelihood that the reports will be disseminated. More publicity provides the basis for nongovernmental actors to pressure governments to comply, thereby reducing their incentives to report. Secretariats are caught in a bind; reports often provide the only basis for identifying noncompliance and creating pressures against it, but using them for that purpose reduces the incentives actors have to supply those reports. In contrast, nongovernmental actors often have incentives to collect and disseminate treaty-related information, but such efforts often lack the same credibility and legitimacy as secretariat reports. More attention needs to be paid to overcoming obstacles to the dissemination of reports, a process that is frequently posited as essential to inducing treaty compliance.

Policy Prescriptions

Having established that some treaty rules elicit more compliance than others and having provided insights regarding the sources of treaties' influence on behavior, we can now ask the question, "What can governments do to increase compliance?" This section uses the analytic framework of the primary rule system, the compliance information system, and the noncompliance response system presented in chapter 2 and the empirical work of later chapters to identify three general propositions and specific subsidiary recommendations about how policymakers can design compliance systems to elicit higher levels of compliance.

Devising an effective compliance system requires that parties select among alternative policies based on a comparison of their success in placing actors within the incentives-ability-authority triangle. Although the three compliance system components are discussed separately below, they are very interdependent. The choice of primary rules has important implications for how easily actors can monitor compliance and how likely actors will be to respond to violations. If funding mechanisms rather than sanctions are to be used to respond to violations, then self-reporting may improve since actors seeking funding will have incentives to report. Although a perfect compliance system cannot be designed, even a good system requires refinement and adaptation to particular environmental problems, the actors involved, and the larger political and social context. The problem of inducing compliance can be managed but not solved. At the same time, for the problem to be managed well requires that all three aspects of the compliance system be dealt with in an integrated and coherent manner, recognizing the impacts that changes in one subsystem will have on the ability to succeed in the other.

Devising Opportunistic Primary Rule Systems

Experience with intentional oil pollution highlights the importance of assigning treaty responsibilities to those actors that are likely to fulfill them. Negotiators often have opportunities to pursue the same environmental goal through quite different primary rules. Although much negotiation surrounds questions regarding what the treaty's goals should be, much negotiation also surrounds questions regarding how to achieve the goals nations agree to pursue. As the shift from regulating tanker operators to tanker owners demonstrates, actors with equal potential to benefit the environment if they comply may vary considerably in terms of how disposed they are to comply or how susceptible they are to monitoring and enforcement. Two rules that would achieve the same result if compliance were perfect can regulate actors with quite different likelihoods of complying.

Negotiators designing primary rules should take advantage of two facts. First, some actors prove more likely to comply with a given rule than others. Not only are some governments and private actors more compliant in general, but patterns of likely compliance vary from issue to issue. Second, different rules impose different obligations on different actors. Although they are framed in universal language, the rules of environmental treaties regulate activities that are rarely uniformly distributed. States' levels of general and issue-specific environmental concern, economic interests, and international power certainly account for much of the variance in compliance among countries. However, these factors do not uniquely determine the rules that nations will adopt to address a given environmental problem.

Frame primary rules to match patterns of compliance burden to patterns of expected compliance. This recommendation simply restates the need for primary rules to ensure that governments and private actors have the incentives, ability, and authority they need to comply with treaty rules.

If power and interests underdetermine the form of a treaty's rules, then within a range of politically viable options some rules may match compliance burdens to expected compliance better than others. Desired behavioral change can be achieved by taking advantage of opportunities provided by the variance in likely compliance rates at the levels of governmental and private actors.[12] Two rules directed at the same environmental goal may impose burdens on different sets of states, one set of which may prove more likely to comply or enforce than the other. Defining primary rules so as to place burdens on the former provides a path of least resistance for creating environmental treaty rules.

Frequently most treaty activity is concentrated in a few states. For example, for thirty years six countries have accounted for over 50 percent of all tanker registries, and fourteen have accounted for 75 percent.[13] Some of these states are flag-of-convenience states that in many cases have shown little interest in oil pollution regulation. In the area of climate change, China, the former Soviet Union, and the United States account for over 50 percent of carbon dioxide emissions due to energy production, and current estimates suggest that China and India account for roughly half the methane released from rice cultivation.[14] Using level of development as a proxy variable for likelihood of compliance suggests that the decision to address climate change by restricting carbon dioxide emissions, methane emissions, or both will have major implications for how well and how quickly member nations will comply with any agreement.

Regulate actors already subject to regulation. Such actors may prove more likely to comply with treaty provisions than others: they are better informed of regulations, are subject to established monitoring and

12. A good example from domestic regulation of matching burdens to likely compliance involves the policy of mandating that employers withhold income taxes throughout the year rather than requiring employees to provide lump-sum payments annually.

13. Lloyd's Register of Shipping, *Statistical Tables* (London: Lloyd's Register of Shipping, 1990); and R. Michael M'Gonigle and Mark W. Zacher, *Pollution, Politics and International Law: Tankers at Sea* (Berkeley: University of California Press, 1979), 56.

14. Susan Subak, Paul Raskin, and David Von Hippel, *National Greenhouse Gas Accounts: Current Anthropogenic Sources and Sinks* (Boston: Stockholm Environment Institute, 1992), 17 and 26.

enforcement systems, and may have a culture or habit of compliance. The oil pollution case illustrates this point at the corporate and individual levels. Tanker construction had been subject to international safety and load line standards long before MARPOL requirements came into effect. A large, well-oiled regulatory infrastructure already existed. A culture of compliance with such standards had become established. Classification societies readily incorporated new equipment requirements into existing construction standards. Although fears of being barred from port or detained for violations undoubtedly played large roles in inducing compliance, many tanker owners built ships with SBT because "those were the rules."[15] Oil and shipping companies initially opposed these requirements, but once they had been promulgated, generally did not conduct cost-benefit analyses to decide whether to comply. Rather, especially given the involvement of other actors in the transaction, a buyer simply had a new tanker built to current specifications. Certainly absent explicit directions, after 1979 a shipbuilder would build a tanker with SBT. Legal considerations were part and parcel of ship design procedures.

This pattern of compliance was not the case for the discharge standards. Prior to 1954, discharge procedures were subject to little if any regulation in most countries.[16] Although some oil companies developed policies to restrict discharges, most captains discharged ballast and tank cleanings based on operational ease and economic costs rather than legal considerations.[17] No culture or habit of compliance developed. Even informing captains of discharge rules was more difficult than informing the relatively few shipbuilders and owners who were already linked to a construction standards information network.

15. Interview, Ernest Corrado, American Institute of Merchant Shipping, Washington, DC, 8 April 1992.

16. For examples of some early national regulations on oil pollution, see Sonia Zaide Pritchard, *Oil Pollution Control* (London: Croom Helm, 1987), 25–30.

17. See, for example, Mobil's policy as described in W. M. Kluss, "Prevention of Sea Pollution in Normal Tanker Operations," in Peter Hepple, ed., *Pollution Prevention* (London: Institute of Petroleum, 1968), 113. See also G. Boos, "Revision of the International Convention on Oil Pollution," in *International Conference on Oil Pollution of the Sea* (Rome, Italy: 1968); and the voluntary agreement undertaken by the oil companies during the 1920s and 1930s, which is described at length in Pritchard, *Oil Pollution Control.*

This discussion suggests several questions that need to be asked when analyzing the primary rules of existing treaties or devising them for future treaties. The first major question is, "What incentives do the regulated economic actors have with respect to a specified rule?" Will they be likely to comply with laws if their governments implement them? In dealing with pollution problems, the answers depend on the availability and costs of altering inputs or production processes, the ability to pass on costs to customers, the competitiveness and concentration of the industry, the degree of enmeshment in existing informational and regulatory infra- structures, and the value placed on being "green." Another major ques- tion is, "What are the incentives and priorities of the governments of the countries where most of the responsible economic actors are found?" How likely are they to impose regulatory controls in line with the international treaty, and how likely is it that they can implement and enforce them successfully? Factors influencing how these questions will be answered include the nation's degree of development and regulatory capacity, the strength of domestic environmental lobbies and countervail- ing lobbyists for the economic sector to be regulated, and the position of environmental concern among the government's other priorities.[18]

Select primary rules that ease the monitoring burden actors will face. The primary rule system not only determines how compliance burdens mesh with compliance incentives, but also provides the foundation for and inherent constraints upon the compliance information system and the noncompliance response system. The control of intentional oil pollution highlights how transparency—which includes both reporting and verification—is a "function of the way in which behavioral prescriptions are formulated."[19] The legal definition of compliance alters the ease of identifying noncompliance, and hence the likelihood that noncompliance will be identified. Limits that are defined to correspond with actors' capacities to self-monitor and independently verify will elicit higher levels of compliance than others. The instruments needed to monitor discharges at the 100 parts per million (ppm) standard established in 1954 were not

18. See, for example, Philipp M. Hildebrand, "Toward a Theory of Compliance in International Environmental Politics," paper presented at the annual meeting of the International Studies Association, March 1992.

19. Young, "The Effectiveness of International Institutions," 177.

developed until the 1980s.[20] The clean ballast provision of 1969 improved transparency relative to the 100 ppm standard, but failed to address the problem of linking detected violations to responsible violators at a level of confidence sufficient to convince domestic courts, let alone those of flag states. Although it was hoped that the definition would make photographic evidence unimpeachable, international jurisprudence and the sanctity of nations' control over their legal systems has prevented this from becoming the case.[21] Clearly rules that define compliance so as to match existing capacities to monitor and verify compliance are necessary, but not sufficient to improve compliance.

Primary rules also influence the likelihood of monitoring by dictating the resources actors will need to expend to collect compliance information. The framing of primary rules can reduce the resources required to monitor compliance by targeting behaviors that involve fewer actors or actions that need to be monitored, are more transparent, coincide with activities already subject to monitoring, and involve transactions between actors rather than actions over which a single entity has control.[22] The Montreal Protocol's focus on the few chlorofluorocarbon producers rather than the myriad consumers provides a clear case of using the first strategy. The switch to equipment regulations in the control of oil pollution demonstrates all four factors that contributed to the dramatically better monitoring of equipment standards than discharge standards: tanker new builds are less frequent than tanker voyages; equipment is

20. Proof that the 100 ppm standard was "functionally unenforceable" was presented to the 1962 conference (M'Gonigle and Zacher, *Pollution, Politics, and International Law,* 225). By 1976 "no oil content meters capable of reliably measuring the 100 ppm standard" had yet been developed; see MEPC VI/4 (30 Sep 1976), 2. As of 1980, "a reliable monitoring system to inform a ship's crew when the oil content of a discharge exceeds a permitted maximum ppm figure [did] not exist. And visual inspection of the outflow is an unreliable means of monitoring the oil-water concentration"; see William G. Waters, Trevor D. Heaver, and T. Verrier, *Oil Pollution from Tanker Operations: Causes, Costs, Controls* (Vancouver, BC: Center for Transportation Studies, 1980), 121.

21. Ton IJlstra, "Enforcement of MARPOL: Deficient or Impossible," *Marine Pollution Bulletin* 20 (December 1989).

22. On the strategy of "piggybacking" on existing compliance information systems, see Thomas A. Barthold, "Issues in the Design of Environmental Excise Taxes," unpublished paper, Cambridge, MA, 1991.

far easier to detect than a discharge; classification societies already monitored tanker construction; and buyers, builders, and classification societies all control a tanker's construction. The last of these also provided the foundation for a coerced compliance system that could prevent rather than merely deter violations.

Frame the primary rules in specific terms that identify what actors must take what actions. Frequently vagueness and ambiguity are artifacts of insurmountable differences in negotiating positions. However, treaty rules that leave either the required actions or the responsible actors unclear provide actors with rationales, if not reasons, for noncompliance. Vague rules may induce behavioral change by less instrumental pathways than those that have been the focus of this study. In the case of MARPOL, the treaty's requirement that governments "ensure the provision" of "adequate" reception facilities papered over the failure to resolve the debate about whether governments or industry should provide them and partially accounts for the continuing absence of reception facilities in many parts of the world.[23] Even IMO guidelines regarding adequacy leave individual governments to determine whether a given port requires a reception facility and its capacity,[24] giving actors with vested interests the power to decide whether they need to comply with the rule and making responses to noncompliance by others less likely since they cannot identify what behavior to monitor or whether observed behavior constitutes noncompliance.

The foreknowledge that vague definitions hinder the collection of compliance data reduces the incentives to undertake such efforts in the first place. At the extreme, of course, it is impossible to know whether governments are complying with the hortatory requirements of the Framework Convention on Climate Change that parties "promote and cooperate" in the implementation of policies "aimed at" stabilizing carbon dioxide emissions or the requirements of the Ramsar Convention

23. P. G. Sadler and J. King, "Study on Mechanisms for the Financing of Facilities in Ports for the Reception of Wastes from Ships," MEPC 30/Inf. 32 (12 October 1990) (Cardiff, Wales: University College of Wales, Cardiff, 1990).

24. IMO, *Guidelines on the Provision of Adequate Reception Facilities in Ports: Part I (Oily Wastes)* (London: IMO, 1991 <1976>).

that states make "wise use" of their wetlands.[25] Although it remains to be demonstrated that nonspecific hortatory language can induce behavioral change, such language certainly makes it difficult for conscientious actors to identify and respond to those actors that are not fully committed to achieving a treaty's goals.

Devising Compliance Information Systems to Benefit Information Providers

Once negotiators establish primary rules that facilitate monitoring, effective compliance information systems must establish self-reporting or independent verification systems or both and develop procedures to analyze and disseminate compliance information. Rules amenable to self-reporting and verification will not of themselves result in an effective compliance information system. To succeed, a compliance information system must reduce the demands on those with the ability to collect information on compliance and violation while encouraging them to report that information by facilitating the achievement of their immediate goals, providing them with a return on their investment. If the system relies on self-reporting, the consolidation, analysis, and dissemination of information must assure those reporting that they will not be sanctioned for noncompliance and that the system will generate information not otherwise available that will benefit the information provider. If the system relies on independent verification, it must provide conduits for information from those with incentives to collect it while assuring them that identifying noncompliance will not evoke retaliation and that the information provided will lead to a response.

Design compliance information systems to reduce the resources needed to effectively collect and report information, especially including computerization. As discussed above, the primary rules play a crucial role in determining what activities must be monitored. However, the treaty can also provide the legal authority necessary to allow such monitoring and can assist in the monitoring. MARPOL provided port states with the

25. See Article IV, *Framework Convention on Climate Change*, UN Document A/AC.237/18 (Part II)/Add. 1 (15 May 1992). See also *Convention on Wetlands of International Importance Especially as Waterfowl Habitat*, 2 February 1971, 996 U.N.T.S. 243, reprinted in 11 I.L.M. 969.

authority they needed to inspect ships and authorized delegation of the authority for initial and periodic tanker surveys to classification societies, greatly reducing the burdens on governments. The computerized MOU system helped maritime authorities distinguish likely violators from likely compliers by informing them of other countries' recent inspection results or the absence of such inspections.

After compliance information has been collected, the self-reporting systems examined—IMO's systems for reports on reception facilities, IMO's enforcement reporting system, the MOU's enforcement reporting system, and the oil record book—demonstrate the need for such systems to facilitate reporting of information, to generate information that can be used to assess compliance, and to analyze and disseminate the information provided so that there can be a response, whether direct or diffuse, whether inducements or sanctions.[26]

Establish self-reporting systems that consolidate and disseminate information in ways that further the reporting entities' own policy goals. Although IMO developed systems for self-reporting on available reception facilities and for monitoring inadequate facilities provided by others, the former has been markedly more successful. Similarly, MOU enforcement reporting has been far more frequent than reporting in the IMO enforcement reporting system, although both systems have required essentially the same information. In both successful systems, the responsible secretariat took the data provided and processed and disseminated them to make them useful to the reporting agents. IMO's publication of data on available reception facilities made their use more likely, thus fostering the providing countries' desire to avoid oil discharges off their coasts. The MOU's consolidation of daily reports into a real-time data base that allowed authorities to focus inspections on likely violators gave those same authorities strong incentives to conscientiously report their inspection results. In contrast, IMO has made little use of its enforcement reports or reports identifying states' inadequate reception facilities, giving states few incentives to supply such reports.

26. It is worth noting that numerous difficulties hinder the evaluation of whether self-reporting actually elicits more of the activity being reported on. For our purposes here, it suffices to note that compliance is unlikely to decrease due to improvements in the reporting system.

Furthering the reporting entities' interests also requires that the treaty secretariat avoid attempts to sanction those reporting their own noncompliance. In a self-reporting system the goal of inducing actors to report conflicts with the role of the compliance information system in responding to noncompliance. Neither the MOU enforcement reporting system nor IMO's reception facility system has used reported information to identify noncompliance with substantive provisions. Those designing self-reporting systems must accept that they prove better for evaluating treaty effectiveness than for increasing it by providing grounds for sanctioning. Self-reporting of noncompliance that is linked to positive responses, as in the International Monetary Fund or World Bank systems, has no parallels in the oil pollution case, but does seem a likely means of successfully inducing reporting.

Design self-reporting systems to be as user-friendly as possible. While keeping the same reporting requirement, in 1985 IMO identified clear and standardized formats for information reporting and thereby improved report quality and content at the same time that it increased the number of countries reporting. The MOU reporting system's requirement of daily inspection reporting worked precisely because it was incorporated into the standard operating procedures of the bureaucracy responsible for inspections.[27] Since the MOU was an agreement between maritime authorities rather than foreign ministries, those who negotiated the agreement also established the inspection and inspection reporting procedures that resulted in regular reporting.

Design independent verification systems to foster the involvement of actors with independent incentives and capacities to collect such information. The many obstacles to effective self-reporting, as demonstrated by the oil record book or the IMO enforcement reports, show the need for independent verification of noncompliance, usually by actors bearing the direct costs of the noncompliance. Governments that value environmental protection have been the ones to undertake aerial surveillance and rigorous tanker inspection programs. These governments perceive discharge and equipment violations as costly, monitoring tankers near their

27. On the role of standard operating procedures, see Graham T. Allison, *Essence of Decision: Explaining the Cuban Missile Crisis* (Boston: Little, Brown, 1971).

own shores or in their own ports to protect their own shores from pollution. Yet neither unilaterally nor jointly have they established a global monitoring system—via satellite, for example—of open ocean discharges.[28] The total discharge limit was never monitored because the treaty drafters ignored the fact that flag states, oil-exporting states, and oil companies all lacked incentives to undertake such inspections. Oil companies, which saw LOT as a means of reducing waste of the cargoes they owned, collected data on tankers' use of LOT during the 1970s, but did not release the information to governments for their use in further prosecuting the tankers they owned and chartered.[29] Similarly, governments generally fail to report inadequate reception facilities in other states' ports because these violations pose no threat to the reporting countries' shores.

Nongovernmental actors play crucial roles in such systems. Through the International Oil Pollution Prevention (IOPP) certificate process, classification societies have proved reliable sources of information on equipment compliance, largely because they face major costs to their business if governments or insurance companies discover that information regarding a specified ship is false. Surprisingly, however, IMO has not taken advantage of private entities that collect and publish data on equipment installations to evaluate compliance or treaty influence.[30] The shipping industry has reported on inadequate reception facilities because tanker operators face considerable additional costs if reception facilities are not available and adequate to allow tankers to avoid long delays when they are used. Although the IMO system for verification of inadequate reception facilities by governments has failed, data on inadequate facilities

28. Such a system has been used by the United States to monitor fisheries' treaty violations, however; see Hilary F. French, *After the Earth Summit: The Future of Environmental Governance,* Worldwatch Paper 107 (Washington, DC: Worldwatch Institute, 1992), 30.

29. William Gray, "Testimony," in U.S. House of Representatives, Committee on Government Operations, *Oil Tanker Pollution—Hearings: July 18 and 19, 1978* (95th Congress, 2nd session) (Washington, DC: GPO, 1978).

30. See, for example, Clarkson Research Studies, Ltd., *The Tanker Register* (London: Clarkson Research Studies, Ltd., 1990); Drewry Shipping Consultants, Ltd., *Tanker Regulations: Implications for the Market* (London: Drewry Shipping Consultants, Ltd., 1991); and Lloyd's Register of Shipping, *Register of Ships* (London: Lloyd's Register of Shipping, 1991).

have reached IMO through the International Chamber of Shipping's anonymous questionnaires. Anonymity is essential to eliminate the threat of retaliation that discourages such reporting through the IMO system, in which the reporting country must specify the reporting ship's name.[31] Providing channels for nongovernmental actors with the incentives and capacity to detect violations, as IMO did by allowing governments to delegate IOPP certificate issuance to classification societies, improves the likelihood that violation will be detected when it occurs. NGOs have also contributed to the compliance information system by collecting, analyzing, and disseminating studies of compliance and enforcement more frequently than have the IMO Secretariat or member governments.

Remove any legal barriers to independent verification. Although oil-loading states lacked the incentives to inspect total discharge violations even had they had such incentives, legal barriers prevented such inspections until MARPOL entered into force. Similarly, the right to verify the claims of the IOPP certificate by inspection was essential for states to successfully detect ships violating the equipment standards and to provide incentives for accurate certification by classification societies. By removing the legal barriers to such inspections, MARPOL increased the use of such inspections. Although the new rights did not lead states uninterested in enforcement to begin inspection programs, they significantly increased the effectiveness of the programs that activist states already had in place. As in arms control treaties, new verification rights need to be created and old barriers to verification removed if verification is to actually be performed.

Creating Noncompliance Response Systems That Remove Barriers to Responses

A treaty can create pressures and incentives to induce reluctant actors to respond to noncompliance by others. Alternatively it can identify and

---

31. Since nonstate actors serving as monitors seek to get governments or other actors to sanction a violator, the compliance information system must reduce disincentives to verification posed by the threat of retaliation in cases in which anonymity is not guaranteed. Although anonymity may increase the likelihood of false reports, if followed by independent follow-up by the secretariat or national governments, the desire to have future claims evaluated and independently validated by the secretariat would prevent sanctioning in cases of false reports.

remove barriers that restrain those actors that are already inclined to respond. In the realm of oil pollution regulation, the former has proved consistently unsuccessful, while the latter has had some success. Those lacking exogenous incentives to respond to noncompliance were not influenced by OILPOL or MARPOL's legal obligations to do so. Whether response involved sanctioning of noncompliance by tankers or funding to finance compliance with reception facility requirements, those actors that lacked preexisting incentives to respond have not done so. To say that international law cannot force those without such incentives to respond is not, however, equivalent to saying that international law cannot make responses from those with such incentives more likely.

Remove legal obstacles to the use of effective sanctions by states and nongovernmental actors with existing enforcement incentives. States do not appear to take seriously legal obligations to sanction violations—obligations common to many environmental treaties. Even among states committed to enforcement, numerous calls to increase penalties for discharge violations failed, even while providing these same states with the right to detain ships led some to impose this penalty, which was far more costly to the tanker operator. Making obligations of existing, but unexercised rights had little impact; without mechanisms to sanction governments failing to enforce a treaty, converting a right into an obligation does not increase incentives enough to alter behavior. In contrast, states do take seriously international legal prohibitions against certain sanctions. The unwillingness of port states to detain ships before MARPOL entered into force affirms the restraining influence of international rules. Although it involves domestic law rather than treaty compliance, the United States-Mexico tuna-dolphin dispute in the General Agreement on Tariffs and Trade also illustrates how legal barriers limit the options of environmentally activist states. Removing these legal barriers often requires negotiating redefinitions of the boundaries and definitions of sovereignty. The new right of port states to inspect and detain tankers decreased the sovereign rights of flag states. Without fundamentally threatening the structure of the international system or current core notions of sovereignty, minor modifications can significantly improve enforcement in a given issue area.

Procedures can also provide a role in treaty implementation to NGOs that have incentives to respond to noncompliance. The necessary involvement of classification societies and insurance companies in international oil transportation and the specific authorization of the former in IOPP certificate inspections carried with them the threat that noncompliance would prevent a tanker from receiving the classification, certification, and insurance necessary to operate on world markets. In other issue areas, as in whaling and human rights, NGOs often play an active role, using publicity campaigns, boycotts, and stronger measures against what they perceive as violations, although treaties rarely authorize such actions.

Reduce the frequency with which responses to noncompliance must be made. Violations of discharge provisions could be committed on every tanker voyage, and the success of the attendant deterrence-based regulatory strategy required that a large fraction of these pay large penalties. These requirements proved difficult to meet. In contrast, equipment violations could be committed only once in a tanker's lifetime, representing a dramatic decrease in the number of potential violations. More important, the equipment standards relied on a strategy that largely prevented violations rather than deterring violators from committing other violations later. This strategy effectively elicited initial compliance, making actual violations that required response even less frequent.

Authorize noncompliance responses by one government against nationals of other states. In contrast to engaging in intergovernmental sanctioning, states have proved quite willing to sanction foreign tankers for noncompliance with equipment and discharge provisions. Many governments have used domestic legal proceedings, detentions, and barring from entry to sanction violations by specific tankers. Indeed in the oil pollution context enforcement usually refers to government sanctions against tankers, not government sanctions against other governments. MARPOL expanded the legal authority for such sanctioning through its detention provisions. Many states fail to enforce treaty provisions at all, and of those that do most fail to impose significant penalties or detain ships.[32] However, states certainly prove less reluctant to undertake extensive and

32. See Gerard Peet, *Operational Discharges from Ships: An Evaluation of the Application of the Discharge Provisions of the MARPOL Convention by Its Contracting Parties* (Amsterdam: AIDEnvironment, 15 January 1992).

effective enforcement programs when sanctions can be directed at private companies, whether belonging to nationals of their own or other countries. Authorizing such sanctions increases their frequency by skirting thorny issues of sovereignty and diplomatic discomfort that are raised when sanctions must be directed at other governments.

Design the response system to emphasize responses appropriate to the likely sources of compliance. The framework developed in chapter 2 stressed the need to distinguish noncompliance due to inadvertence and incapacity from intentional violations. The noncompliance response system must identify which of these is the most likely and design appropriate responses. Proposals for positive responses to noncompliance, e.g., the compliance financing mechanisms of the Montreal Protocol, assume that noncompliance arises from incapacity, not intention. The oil pollution treaty provides little insight into how effective such programs are since, despite frequent proposals, they have not been used to address compliance with discharge, equipment, or reception facility requirements.[33] Yet this fact suggests the likely obstacles to such programs: developed states often fail to supply the resources needed to create the incentives and capacity for developing states to meet their environmental treaty commitments. The evidence here is confirmed by the difficulty in negotiating even the limited financial mechanisms of the Montreal Protocol. Even if such programs prove effective when implemented, disputes over the major cause of noncompliance, insufficient concern in developed states, and other disincentives to providing financing make implementation unlikely.

If financing is an appropriate but rarely provided response to incapacity, equal problems obstruct sanctioning efforts. Although sanctioning seems an appropriate response to intentional noncompliance by states, it has also been rarely undertaken. The only significant area in which

33. The lack of financing for reception facilities in developing countries has frequently been noted (see chapter 6). Early versions of Agenda 21 called for developed countries to provide $80 million per year for oil reception facilities in developing countries; see Preparatory Committee for the United Nations Conference on Environment and Development, *Protection of Oceans, All Kinds of Seas Including Enclosed and Semi-enclosed Seas, Coastal Areas and the Protection, Rational Use and Development of Their Living Resources* U.N. Doc. A/Conf. 151/PC/100/Add. 21 (New York: United Nations, 1991).

governments have raised the issue of noncompliance by other states has been with respect to flag states' prosecution of port states' referrals.[34] Government-to-government sanctions beyond those of voicing diplomatic concerns have rarely been attempted. With respect to providing reception facilities, establishing strict penalties, or implementing vigorous port state controls, few efforts have been made to even identify, let alone shame, states not complying. This fact supports the notion that states prove reluctant to use stringent sanctions, such as trade embargoes, to enforce international agreements.[35] Indeed the oil pollution experience provides little evidence that states were even willing to use shaming or jawboning as a response to noncompliance.[36]

Governments must also have sanctions available that fit the crime but involve low costs to the government. The contrast between using penalties and detentions to elicit compliance by tanker operators demonstrates how punishments sufficient to deter violators from committing other violations may not be imposed because they do not fit the crime. Judiciaries usually impose penalties proportional to the perceived harm of the violations, largely independent of calls for deterrence in international treaties. Deterrence proves a desirable side benefit, not a determinant, of the size of a penalty. Fines deemed appropriate to discharge violations have been so low as to provide little if any deterrent value. Detention, in contrast, has been perceived as an appropriate response to a given tanker's equipment violations, but has also deterred such violations.

Reducing the costs of sanctioning to the sanctioning actor proves crucial. MARPOL replaced the huge legal burden involved in investigating discharge violations and prosecuting and convicting violators with the far less costly administrative alternatives of detention or barring from entry, removing the major disincentive to sanctioning, namely the

34. For example, see the French complaints outlined in MEPC 21/16/3 (30 January 1985).

35. The U.S. threat of exclusions from fisheries to induce foreign whaling fleets to make behavioral changes provides an example of when a state has used such techniques; see Dean M. Wilkinson, "The Use of Domestic Measures to Enforce International Whaling Agreements: A Critical Perspective," *Denver Journal of International Law and Policy* 17 (Winter 1989).

36. Abram Chayes and Antonia Chayes, "On Compliance," *International Organization* 47 (Spring 1993), 188–192.

difficulty and cost to the sanctioning government. Detention was therefore more likely than prosecution of a discharge violation, and it was also more costly to the operator. Sanctions that prevent violators from concluding the business enterprise that produces pollution impose greater costs than those that impose fines that can often be considered a bearable "cost of doing business."[37]

As with compliance and monitoring, treaty commitments succeed if they ensure that the burdens for responding to noncompliance fall upon those nations and those actors that have the practical ability, the legal authority, and, most important, the incentives to fulfill them. Especially in the noncompliance response system, the number of actors disposed to respond to noncompliance is likely to be low. Treaty rules can increase responses most by removing any existing legal barriers that inhibit the imposition of effective and appropriate responses.

Contextual Factors and Limits to Generalizability

This book has argued that rules regulating intentional oil pollution have influenced behavior and increased compliance. Even a reader convinced by this argument may well ask, "Since all these cases occurred within the context of a single environmental problem, can these findings be generalized?" Do the findings transfer to other issues? Is the constellation of factors that led to success in the case of oil pollution unlikely to be present in dealing with other issues?[38] The two preceding sections of this chapter have assumed that the oil pollution experience held insights for application to other environmental issues; this section explicitly evaluates that contention.

37. The towing of cars and the use of "Denver boots" rather than parking tickets to deter drivers from committing parking violations suggests a domestic parallel to this.

38. The book's primary goal has been to demonstrate that treaty rules could influence behavior. Holding context constant, studying several cases within a single treaty eliminated variance in several otherwise confounding variables. Generating propositions regarding traits accounting for variance in compliance levels between successful rules and others has been an important but subsidiary goal that has not been equally well served by the methodology selected. To truly address the generalizability argument, research must test these hypotheses in other environmental issue areas.

The oil pollution issue was selected for study precisely because contextual conditions made it unlikely that treaty rules would succeed. Most of the exogenous factors influencing the behavior of governments, tanker owners, and tanker operators with respect to oil pollution warranted against MARPOL's provisions' having much effect. Indeed the general failure of most OILPOL provisions affirms this expectation. Success in inducing behavioral change regarding oil pollution was unlikely because it required costly and immediate changes by a well-organized and powerful industry coupled with relatively small, uncertain, and widely dispersed benefits accruing to unorganized publics. Although the oil and shipping industries are dominated by a few large actors, they certainly do not dominate their realms more than the chlorofluorocarbon, pesticide, power production, or whaling industries. The oil transportation industry includes many small independents that create a highly competitive market. This competition generates ongoing pressures to circumvent regulations that increase the costs of doing business, pressures that are evident in ongoing opposition to efforts to apply equipment requirements to more tankers. MARPOL did not merely reflect states' existing environmental preferences: several governments strongly opposed and voted against the equipment regulations adopted. Whatever incentives governments had to enforce the regulations also militated against the enforcement of the equipment standards, which provided nonexcludable public benefits, and for the enforcement of the discharge standards.

In addition, oil pollution has characteristics similar to many other environmental problems. Reducing oil pollution requires collaboration between states, not merely harmonization or coordination. Companies had continuing incentives to violate MARPOL's provisions, and governments had continuing incentives not to enforce them even after it was signed, as is attested to by the violations of the discharge standards, the frequent nonenforcement, and the lack of SBT equipment on tankers not required to have it. Overcoming these incentives to free-ride constitutes the *sine qua non* of international environmental problems, from climate change and ozone depletion to endangered species, deforestation, and habitat loss.

Having said this, the problem more closely resembles—and the recommendations made here can more likely be generalized to—pollution problems involving externalities than it does wildlife and habitat preservation. Production process changes can often resolve pollution externality problems without threatening the economic viability of the responsible industry. The international conventions addressing acid precipitation, ozone depletion, climate change, river and nonoil marine pollution, and hazardous wastes have all faced choices similar to those made here between operational limits and equipment standards. Altering oil transportation processes is certainly not inherently less costly than making the required changes in many of these areas. In contrast, preserving wildlife and habitats often pits value systems that are more directly at odds against each other, with environmental progress requiring at least temporary cessation or limitation of the economic activity responsible. Although countries worked for years to ban commercial whaling, banning oil transportation by sea has neither been necessary nor been considered.

Oil transportation involves more interaction between national governments and foreign polluters, thus making the latter less capable of using political pressure to avoid sanctions. This is not unique to oil pollution, however. Due to the internationalization of the global economy, many environmental problems can be attacked through trade-related measures. Conventions addressing endangered species, hazardous wastes, whaling, the cutting of tropical timber, and ozone-depleting substances all include articles involving trade as a sanction. Even if trade-related measures are inadequate to resolve the problem, the oil pollution experience demonstrates that they do provide a means by which activist governments can influence the behavior of foreign nationals. Indeed prior to MARPOL international law prevented governments from sanctioning foreign tankers in most instances of violation; part of MARPOL's success stems precisely from its adoption of detention rights that skirted legal sovereignty issues to allow such sanctioning. When harmful practices are practically and legally in the exclusive jurisdiction of a single state, the MARPOL experience suggests that at least the latter may be open to some modification.

Some of this book's findings and recommendations undoubtedly apply more widely and accurately than others. Almost all environmental treaties include requirements for reporting on enforcement, compliance, and other treaty-related activities, and the hypotheses generated here do not appear to be especially unique to or constrained by the characteristics of oil pollution as an issue or the oil market. Attempts to induce

governments to impose higher fines that rely exclusively on legal requirements without supporting review processes seem unlikely to be any more
successful in other treaties than they were in OILPOL. Likewise, the
failure of new inspection and enforcement obligations (as opposed to new
rights) to induce reluctant governments to monitor and enforce seems
unlikely to be the result of factors unique to oil pollution. In other
instances the experience with oil pollution may hold even more strongly.
For example, the funding needs in the Montreal Protocol, let alone the
climate change agreement, swamp those in the requirements for reception
facilities; developed nations may show a stronger commitment to providing funds to avert ozone depletion and global warming, but the complete
failure to establish resource transfers for oil waste reception facilities does
not bode well.

Some of the successes in regulating oil pollution also are either difficult
to duplicate in other issue areas or appear to be conditioned by relatively
unique features of the oil pollution problem. An effort that seeks robust
causal conclusions and strong internal validity by focusing on one issue
area necessarily sacrifices some ability to generalize to other areas. Looking at changes in rules within a single issue area allowed for significant
control of variation in other causal variables. By definition, however, this
makes it impossible to know the degree to which the success of certain
rules was conditioned by contextual features of the issue area studied.
Several of these factors nonetheless appear sufficiently likely to have
conditioned the successes and failures of certain rules that they deserve
mention.

I have noted that how who gets regulated influences the likely level of
compliance; this proposition rests on evidence from two industries—
tanker operations and tanker construction. Compliance may prove far
harder to elicit in cases in which numerous otherwise unregulated individuals have caused the environmental problem, as with deforestation,
wetlands degradation, and endangered species loss. When the detrimental
activities are undertaken exclusively within a single country's borders, as
with fossil fuel use and livestock and rice cultivation for climate change
or the preservation of world heritage sites or biodiversity, improving
compliance requires two-level enforcement. Concerned governments must
sanction reluctant governments to induce them to enforce treaty commitments against their national citizens, and such sanctioning proves infre-

quent. The obstacles to effective enforcement in such situations are evident in the low figures for flag state prosecutions and fines. Success has been elusive when governments have been reluctant to fulfill their international commitments rather than merely inattentive to them.

Certainly costs matter. If MARPOL had adopted a more rapid schedule and required all tankers to retrofit with SBT, the substantially higher costs would probably have caused higher levels of noncompliance. Solutions that worked in regulating oil pollution may also prove irrelevant to other problems. Some environmentally harmful activities are not susceptible to technological solutions or quantitative requirements that can be easily monitored. The array of available regulatory strategies depends, at least in part, on features unique to the activities causing the environmental damage.

The level of environmental concern, especially among internationally powerful states, has been a sufficiently important influence on the dynamics of oil pollution regulation and compliance that it deserves separate attention. Strong environmental pressures from the United Kingdom and the United States have been a constant prerequisite for adoption of any of the rules that have been successful in increasing compliance.[39] Where insufficient concern existed to address the underlying source of noncompliance or nonenforcement, rules were either not agreed to or were designed to fail. Treaties do not establish rules that exceed the desires of dominant states. These desires, however, underdetermine observed behaviors. The contrast between simultaneously high levels of compliance with the equipment standards and low levels of compliance with the discharge standards highlights how environmental concern in states like the United States may be necessary, but is not sufficient to ensure successful implementation of a treaty. The depth and breadth of environmental concern among powerful states defines the range of options for rules addressing compliance and enforcement; within this range, however, success is not predetermined, and some rules may elicit more compliance than others.

39. A more extended version of this point is made in Ronald B. Mitchell, "Intentional Oil Pollution of the Oceans," in Peter Haas, Robert O. Keohane, and Marc Levy, eds., *Institutions for the Earth: Sources of Effective International Environmental Protection* (Cambridge, MA: The MIT Press, 1993).

Establishing compliance systems along the lines suggested above would certainly fail to improve compliance unless it were accompanied by support among several dominant states in the international system. Without strong concern, effective compliance systems are unlikely to be devised in any event. The existence of strong concern, however, provides no assurance that successful rules will be adopted nor that adopted rules will be successful. Even countries committed to environmental improvement must work at crafting treaties that will elicit compliance. In environmental arenas in which the level of environmental concern is low, no efforts are likely to produce successful compliance systems. Once such concern arises, however, the recommendations made here may well provide guidance for those seeking to elicit compliance.

Compliance and Effectiveness

The foregoing study prompts the question of whether MARPOL rules, for all their influence on compliance and behavior, have actually solved the oil pollution problem. Has MARPOL succeeded in terms of effectiveness as well as compliance? In Oran Young's terms, has MARPOL had problem-solving as well as behavior-changing impacts?[40] Although I have evaluated compliance in this issue area against explicit treaty provisions, we are most interested in whether treaties produce the environmental improvements that motivate their negotiation in the first place. Compliance and behavioral change are valuable only if they lead to the accomplishment of treaty goals. Evaluating problem-solving effectiveness requires that we make an often subjective choice among various possible definitions of the problem. For one thing, "sharp statements of objectives seldom are achieved" in international environmental treaties, making it difficult to find the yardstick against which to measure effectiveness.[41] Even if one defines effectiveness as the degree of environmental improvement, the multiple causes of most environmental phenomena and the

40. Marc Levy, Gail Osherenko, and Oran Young, "The Effectiveness of International Regimes: A Design for Large-scale Collaborative Research," unpublished draft manuscript (Hanover, NH: Dartmouth College, 4 December 1991).

41. David A. Kay and Harold K. Jacobson, eds., *Environmental Protection: The International Dimension* (Totowa, NJ: Allanheld, Osmun & Co., 1983), 18.

generally poor quality of data make identification of causal links extremely tenuous.[42]

Recognizing these problems does not obviate the need to attempt such analysis, however. With respect to the issues that have been the primary focus of this study, a narrow definition—"reducing intentional oil pollution from tankers"—can expand along one or both of two axes. The problem can be broadened to include reducing all pollutants from tankers or from ships generally. Or it can be broadened to include reducing all sources of marine oil pollution. How effective, if at all, have the international efforts described here been in solving the problem defined in each of these three ways?

Have the OILPOL and MARPOL conventions actually reduced intentional oil pollution from tankers? Many of the data presented in chapter 7 on discharge violations suggest that intentional discharges remain commonplace among some subset of the tanker fleet. Expert estimates of the number of intentional oil discharges have generally declined over the last several decades.[43] Over the same period, the number of spill sightings and bird oilings have remained relatively constant.[44] However, neither expert estimates nor environmental quality data are of sufficient quality to make a strong case that intentional oil discharges have increased or declined over time. A general lack of consistent time series data and difficulties of discriminating intentional slicks from accidental spills and natural seepage are only a few of the problems one faces in evaluating environmental improvement. However, we can easily identify factors that suggest why the agreements we have studied may well not have reduced intentional oil pollution. The decision not to require immediate retrofitting of both SBT and crude oil washing (COW) means that vessels will continue to legally operate with only one of these technologies well into

42. Robert O. Keohane, Peter M. Haas, and Marc A. Levy, "The Effectiveness of International Environmental Institutions," in Haas, Keohane, and Levy, eds., *Institutions for the Earth*, 7. See also Kay and Jacobson, *Environmental Protection*, 320.

43. For a longer discussion of these problems and an evaluation of environmental improvement, see Mitchell, "Intentional Oil Pollution of the Oceans," especially figure 5.1, 187.

44. C. J. Camphuysen, *Beached Bird Surveys in the Netherlands, 1915–1988: Seabird Mortality in the Southern North Sea since the Early Days of Oil Pollution* (Amsterdam: Werkgroep Noordzee, 1989).

the next century. Additionally, tanker operators intent on avoiding the costs of SBT and COW can avoid using these technologies properly. Although such instances are rare, some tanker operators have been caught using their segregated ballast tanks to carry cargo.[45] The absence of reception facilities has already been noted as an obstacle to even conscientious tanker operators' avoiding illegal discharges. Besides, the large increase in the amount of oil transported by sea could easily be leading to an overall increase in the total amount of oil entering the ocean even if each tanker is discharging less oil less often (see figure 3.1). Therefore, even if tanker owners have installed SBT and/or COW as required and even if many tanker operators refrain from making intentional discharges, the intentional oil pollution problem might show few signs of improvement. Intentional discharges surely would have continued to grow as a problem in the absence of the OILPOL and MARPOL regulations. The higher levels of compliance achieved through equipment standards, because of the obstacles and disincentives they raise to discharging oil at sea, has reduced and over time will further reduce intentional oil pollution.

More generally, increased compliance is neither a necessary nor a sufficient condition for effectiveness. A high level of compliance is not necessary. Noncompliance with an ambitious goal may still result in considerable positive behavioral change that may significantly mitigate if not solve an environmental problem. Besides, a high level of compliance is not sufficient. High compliance levels with rules that merely codify existing behavior or rules that reflect political rather than scientific realities will prove inadequate to achieve the hoped-for environmental improvement. For example, compliance with the Montreal Protocol may prove perfect but too late for the earth to avoid irreversible harm from stratospheric ozone loss.[46] However, compliance can provide a valuable proxy for effectiveness, since increased compliance will result in more environmental improvement so long as the rules do not have perverse effects, although the improvement may still be insufficient to mitigate the

45. William P. Coughlin, "Two Ships Barred from Unloading Oil in Boston," *Boston Globe* (1 November 1990).

46. See Edward A. Parson, "Protecting the Ozone Layer," in Haas, Keohane, and Levy, eds., *Institutions for the Earth*.

problem. In most cases negotiated rules have a positive relationship to better management, if not resolution, of the environmental problem.[47] Under these conditions, higher levels of compliance will lead to higher levels of effectiveness, all things being equal. In the case of intentional oil pollution, use of the required equipment would reduce a tanker's discharges to levels equal to or less than those required under the total discharge standard.[48] Under assumptions of perfect compliance, therefore, the two rules would be equally beneficial to the environment. Since the equipment rules have achieved higher levels of compliance, we can safely infer that they have also increased treaty effectiveness. Even if insufficient to solve the intentional oil pollution problem, MARPOL has mitigated it.

Have OILPOL and MARPOL resulted in a reduction in the wide array of potential types of pollution from ships? The regime and institutions initially established to address intentional oil pollution have provided a forum in which other vessel-source pollutants have been addressed. Although the 1954 OILPOL agreement addressed only intentional oil pollution, states did turn to the Intergovernmental Maritime Consultative Organization as the natural forum within which to address accidental oil pollution from tankers after the *Torrey Canyon* disaster in 1967. A number of intergovernmental and private agreements had been reached by 1969 that addressed many of the then-current concerns.[49] This use of IMO continued after the *Exxon Valdez* accident in 1990, with negotiation of the Oil Pollution Response Convention and recent design and equipment amendments to MARPOL directed at accidental spills.[50] IMO has also expanded beyond oil pollution in other directions as well. The London Dumping Convention of 1972 addressed the dumping of wastes

47. John Kambhu has explored the interaction of these two effects at the domestic level in his excellent article, "Regulatory Standards, Noncompliance and Enforcement," *Journal of Regulatory Economics* 1 (June 1989).

48. MEPC 30/Inf. 13 (19 September 1990).

49. *International Convention on Civil Liability for Oil Pollution Damage*, 29 November 1969, reprinted in 9 I.L.M. 45 (1969); *International Convention Relating to Intervention on the High Seas in Case of Oil Pollution Casualties*, 29 November 1969, reprinted in 9 I.L.M. 25 (1969).

50. *International Convention on Oil Pollution Preparedness, Response and Co-operation*, 29 November 1990, reprinted in 30 I.L.M. 735 (1991).

at sea, and four of MARPOL's five annexes dealt with pollution from ships other than oil, namely chemicals, other hazardous substances, sewage, and garbage. Most recently the Marine Environment Protection Committee has also begun addressing various air pollution emissions from ships, such as chlorofluorocarbons and sulfur dioxide.[51] Certainly the regulations developed under IMO's auspices to reduce various pollutants generated by ships have become consistently broader and more stringent over time. To determine whether these rules have elicited compliance or achieved effectiveness would require a study similar in length to the current one. Oil tanker accidents have decreased relatively consistently over time.[52] However, the intentional oil pollution regime specifically and the international maritime regulatory regime more generally certainly have been effective in providing a legitimate and readily available forum that has facilitated discussion and negotiation of controls of various vessel-source pollutants. It seems unlikely that as many regulations would have been negotiated as quickly without the oil pollution experience as background.

Have OILPOL and MARPOL led to reductions in the overall problem of marine oil pollution? One critique of ship-generated oil pollution regulations is that they have not addressed the major source of oil pollution, namely land-based sources. And oil interests have opposed efforts to regulate air pollution from ships because it represents such a small part of the air pollution problem.[53] The failure to address other sources of marine oil pollution can be seen as the result of two forces. At the time that oil pollution was initially addressed, ships were considered—and, indeed, probably were—the major source of marine oil pollution. Framing this issue as reducing pollution from a particular source and negotiating subsequent regulations within an organization whose mandate was maritime regulation precluded negotiators from expanding regulations to other sources. Subsequent IMO history confirms that there has been an unsurprising bias toward the further regulation of ships rather than toward expanding regulations to other sources of a given marine pollutant. Certainly any attempt by IMO to regulate other

51. MEPC 31/WP.13 (4 July 1991), 16–19.

52. "How MARPOL Has Changed," *IMO News,* 1992 (March 1992), 7.

53. See, for example, MEPC 32/12/1 (16 January 1992).

sources would undoubtedly be resisted as illegitimate and inappropriate both within and outside the organization. Although the OILPOL and MARPOL regime can claim little credit for it, other sources of marine pollution have nonetheless been regulated. The United Nation's Environment Programme's Regional Seas agreements have addressed a wide array of marine pollutants from various sources, and specific agreements have been reached to regulate land-based sources of pollution.

Finally, comparison of equipment and discharge standards raises the question of whether rules that have achieved high levels of compliance may be too costly, economically inefficient, and not cost-effective. The argument here does not argue for command and control type regulation, but does argue that evaluating likely compliance levels is an important and relevant consideration in policy debates. The experience with discharge standards demonstrates that a nominally cheaper, more efficient policy may simply not be capable of inducing the level of compliance needed to achieve a socially desirable outcome. In contrast, a more expensive, inefficient policy may prove more enforceable or otherwise more likely to elicit the high levels of compliance needed to achieve this outcome. Before choosing such a policy, the benefits of the socially desired outcome must of course be evaluated to determine whether they warrant the higher costs of compliance. I have not attempted to address whether the largely unquantifiable benefits of reduced oil discharges outweigh the admittedly high costs of installing SBT. If they do not, SBT requirements will prove to have been a poor social choice. However, we cannot rule out such options simply because more efficient options can conceivably achieve the intended environmental end. Policies must be evaluated in terms of compliance as well as efficiency, cost, and equity. When they elicit the compliance necessary to achieve the environmental goal, more efficient solutions are clearly preferable. When compliance with efficient solutions appears to be significantly less likely than with alternatives, we must examine whether the compliance costs of inefficient solutions are nonetheless outweighed by the benefits that high levels of compliance provide.[54]

54. This discussion focuses on the relationship of compliance costs to compliance level. The associated problem of the relationship between enforcement costs and compliance has been extensively addressed in attempting to identify optimal levels of regulatory enforcement. For example, see Gary S. Becker, "Crime and

The Need for Further Research

The research for this book raises as many questions as it has answered. My remaining task is to lay out the more important of these to suggest areas for productive future research while simultaneously noting some of the limits of the current study. The study of international environmental politics in general, and issues of compliance and effectiveness in particular, is currently receiving a large and rapidly growing amount of scholarly and policy attention.

This book's primary goal has been to answer the question, "Do environmental treaty rules influence behavior?" Specifically, this study has used several cases of rules drawn from international regulation of intentional oil pollution to demonstrate that environmental treaty rules can influence behavior. Evaluating these rules and comparing them with others that have had no apparent influence on behavior has allowed us to discover several answers to the question, "Which types of rules make a difference?" Specific criteria for designing treaty compliance systems to enhance compliance have been delineated in the preceding pages. While evaluating behavioral change at a generally high level of aggregation, the study has also identified specific pathways by which the rules have caused the observed changes in behavior.

As with most studies, addressing this array of issues has raised a large set of complementary questions that remain unanswered. The study has not shown that treaty-induced compliance is common. More broadly, the study has not addressed the larger question of whether treaty rules generally prompt changes in behavior. Indeed it has presented several clear cases in which rules failed to influence behavior. Determining if compliance itself, whether treaty-induced or not, occurs frequently in international environmental affairs or international affairs more generally demands study by those evaluating treaty rules and behavior across a wide range of treaties.

Punishment: An Economic Approach," *Journal of Political Economy* 76 (March/April 1968); George J. Stigler, "The Optimum Enforcement of Laws," *Journal of Political Economy* 78 (1970); and Oran Young, *Compliance and Public Authority: A Theory with International Applications* (Baltimore: Johns Hopkins University Press, 1979), chapter 7.

The book's findings regarding the types of rules that can influence behavior are, by necessity, artifacts of the issue area selected for study. Certain compliance strategies simply were not used in addressing intentional oil pollution and therefore could not be evaluated. My claim that reciprocity is rarely used to induce compliance with environmental treaties may be an artifact of the cases chosen for study. The use of intrusive verification schemes in environmental affairs, like those under the whaling treaty and wetlands treaties, has yet to be evaluated. Thus neither OILPOL nor MARPOL established any significant positive incentives directly related to compliance, such as financial and technology transfer programs, yet such approaches may well prove to be more effective in eliciting compliance.[55]

Designing a study to produce unambiguous evidence of a linkage between treaty rules and behavioral change demanded that I ignore less instrumental strategies for compliance. Restricting the book's focus to the impact on behavior of explicit treaty rules risks missing the fact that the compliance systems used to alter behavior are nested among broader norms, principles, and processes that may play far greater roles in altering behavior. "Soft law" involving norms, principles and guidelines, informal agreements, and tacit bargaining strategies may wield considerable influence over behavior. Larger "behavioral alteration" systems, of which treaties and their compliance systems form only a part, may exercise influence not only through altering behavior in an instrumental sense, but by successfully changing the values and interests of the actors involved.

Many theorists contend that treaties alter behavior by leading states to adopt broader and longer-range views of their interests, providing new scientific information that clarifies policies to achieve existing goals, or causing governments to learn new goals.[56] As Alexander Wendt argues,

55. Evaluating the financial mechanisms in various treaties is already the focus of one study; see Center for International Affairs and Center for Science and International Affairs, "Developing Effective Mechanisms for Transferring Financial Resources for Environmental Protection," draft proposal (Cambridge, MA: Harvard University, 14 June 1993).

56. See, for example, Robert O. Keohane, *After Hegemony: Cooperation and Discord in the World Political Economy* (Princeton, NJ: Princeton University Press, 1984); Peter M. Haas, *Saving the Mediterranean: The Politics of Interna-*

"international institutions can transform state identities and interests" and through so doing change their behavior.[57] Other tactics, including education, ongoing discussions and meetings, and technological research may not only achieve the same goals, but may prove both more efficient and more effective. Research is also needed to evaluate treaties that establish exclusively hortatory goals. Such treaties may have significant behavioral impacts, even when "compliance" proves impossible to measure. This book decidedly does not exclude such transformation as an important pathway by which institutions change behavior. It has simply been intended to ask how we can mitigate international environmental problems through the less ambitious means of changing the behavior of nations and their citizens even when it proves impossible to alter their identities and interests.

The same treaty rules also undoubtedly influence different governments, corporations, and individuals through different causal pathways.[58] As a simple example, providing scientific or technical expertise can produce significant changes in the behavior of a developing country while producing no such changes in a more industrialized state. Examining variances in countries' responses to the same treaty rules can identify necessary national-level conditions for treaty success in eliciting compliance and point to interactive effects between treaty and national variables.

Certainly the work here on intentional oil pollution requires comparison to work in other environmental as well as nonenvironmental issue areas. Such work can identify both the degree to which the findings of this study are generalizable and the influence that issue-specific variables have on the success of various approaches to eliciting compliance. As with identification of national-level factors influencing compliance, vari-

---

tional Environmental Cooperation (New York: Columbia University Press, 1990); and Edward A. Parson and William C. Clark, "Learning to Manage Global Environmental Change: A Review of Relevant Theory," unpublished paper, Cambridge, MA, 1991.

57. Alexander Wendt, "Anarchy Is What States Make of It," *International Organization* 46 (Spring 1992), 392. In this book I consciously limit myself to what Wendt terms a "rationalist" view in which institutions "change behavior but not identities and interests."

58. Levy, Osherenko, and Young, "The Effectiveness of International Regimes."

ance across different types of problems can identify the conditions necessary for treaty-induced compliance and interactions between rules and issue areas that inhibit or facilitate compliance. Little research has yet systematically explored the wide variance across different environmental treaties in eliciting compliance and achieving desired behavioral changes. The work here clearly demonstrates that some variance will be due to differences in how the treaties frame their substantive provisions and design their compliance systems. However, at least two other factors certainly play key roles in such variance across issue areas. First, different levels of international concern regarding a problem may have large effects on the level of compliance achieved. Second, some types of activities are inherently more difficult to regulate than others. Some environmental remedies need only alter current production processes to eliminate the component causing the environmental externality, as with oil pollution and acid rain. In other cases achieving treaty objectives may require reducing the level of activity to correspond to the relevant ecosystem's carrying capacity, as with fisheries agreements. In still other problem areas treaties may need to completely ban an activity in certain areas or for certain periods, as with whaling, trading in endangered species, and atmospheric nuclear testing. Several research teams involving many international scholars from a wide array of disciplinary backgrounds are currently probing into these and related issues.[59] If this study contributes to those efforts, it will have achieved its aims.

The previous section on limits to generalizability highlighted various contextual factors, such as political concern in dominant states, that create the conditions necessary for the adoption and implementation of effective environmental treaty rules. Understanding what limits contextual factors place on compliance and what factors determine that context can make it easier to design politically viable yet successful environmental compliance systems. Comparative cross-treaty and cross-country studies are already being conducted by several international research teams.[60]

59. See note 27 in chapter 1.

60. Edith Brown Weiss of Georgetown University Law Center and Harold Jacobson of the University of Michigan are leading an especially broad-based research program evaluating compliance by ten countries with five environmental treaties.

Comparisons across different treaties can highlight the impact of contextual factors on the effectiveness of similar compliance systems or the relative effectiveness of different compliance systems. Factors that are not policy-manipulable in the short run, such as scientific understanding of (and hence governmental concern over) the risks of a particular environmental threat or solutions to it, may be policy-manipulable in the long run. Numerous questions suggest themselves. How do wildlife preservation treaties compare with pollution regulation accords? Do bilateral agreements prove better or worse in eliciting compliance than multilateral agreements? In general, how does the number of parties influence compliance levels?

This research could and should be extended to determine what valid lessons can be drawn from broader international relations research efforts, for example those investigating compliance, verification, and enforcement issues in arms control.[61] Comparing and contrasting the arms control experience with that of environmental treaties could shed light on both areas of study.[62] Similar comparisons might usefully be made with compliance provisions and experience in trade and human rights conventions.[63] The proposed Commission on Sustainable Development for monitoring adherence to the principles in Agenda 21 of the United Nations Conference on Environment and Development (UNCED) was modeled after the U.N. Commission on Human Rights, yet little close study has documented whether and why the human rights experience should easily and successfully transfer to the environmental realm.[64]

61. For example, see John S. Duffield, "International Regimes and Alliance Behavior: Explaining NATO Conventional Force Levels," *International Organization* 46 (Autumn 1992).

62. An initial effort has been made in Jesse Ausubel and David Victor, "Verification of International Environmental Agreements," *Annual Review of Energy and the Environment* 17 (1992).

63. On trade, see Ethan Kapstein, *Governing the Global Economy: International Finance and the State* (Cambridge, MA: Harvard University Press, 1994).

64. Paul Lewis, "Delegates at Earth Summit Plan a Watchdog Agency," *New York Times,* 7 June 1992.

Conclusion

Nations can improve compliance with environmental treaties. This book's detailed analysis of rules regulating intentional oil pollution demonstrates that treaty rules can channel the behavior of both governments and industry into activities conforming with the treaty's proscriptions and prescriptions. In regulating any given activity, the distribution of power, interests, and environmental concern across countries and the traits of the relevant economic sectors place upper limits on but underdetermine the rules that can be agreed to and the degree of compliance possible. By acknowledging these limits and realizing that the same environmental goal may often be achieved by regulating quite different activities, policymakers can improve compliance and benefit the environment by regulating those sectors more likely to comply in those countries more likely to implement and enforce. This matching of patterns of compliance burden with patterns of expected compliance places the choice of prescriptive and proscriptive rules at the center of any effective compliance system. If made well, these choices can be essential determinants of future international behavior toward the environment at both the government and industry levels.

These choices also delimit the ease, likelihood, and effectiveness of efforts to elicit self-reporting on, independent verification of, and responses to noncompliance. Yet even if few options of activities to regulate are available or new primary rules cannot be agreed upon, policymakers can still increase the likelihood that actors will provide information on compliance and that they will respond to noncompliance when it is detected. Careful crafting of the compliance information system and the noncompliance response system can increase the likelihood that noncompliance will be identified. In most cases this involves removing the legal, practical, or institutional barriers that prevent or deter those with incentives to report, monitor, or respond from actually doing so. Rather than by imposing obligations to implement agreements on reluctant actors, success in oil pollution regulation has stemmed from working toward the goals of, removing barriers limiting, and providing rights to those governments and private actors with preexisting incentives to monitor or enforce agreements.

For these findings to prove useful, subsequent research will need to use the lessons learned in these cases to identify ways that other existing treaties can be improved and future treaties can have more compliance "designed in" from the start. This book's greatest value will stem from providing those policymakers devising and refining other agreements with a detailed analysis of the successes and failures experienced in thirty years of attempting to improve compliance with international rules regulating intentional oil pollution. Despite making continuing references to international environmental politics as a new field, many environmental treaties have been in force long enough for us to begin evaluating their behavioral impacts.[65] Although it may take time to develop empirical evidence of compliance, many environmental agreements stand ready to be analyzed by those inclined to find data addressing whether international rules effect compliance.[66]

Whether the nations of the world succeed in averting the many international environmental threats that loom on the horizon will depend not on negotiating agreements to alter the behaviors that harm the air, land, and water, but on ensuring that those agreements succeed in inducing governments, industry, and individuals to change their behavior. We can hope and work for a day when all nations and their citizens will be sufficiently concerned about the environment that we will not need international law to outlaw pollution and dictate environmentally benign behaviors. Until then, however, careful crafting and recrafting of international treaties provides one valuable means of managing the protection of the global environment.

65. For historical examples of claims that international environmental studies are new, see Richard Falk, "Environmental Policy as a World Order Problem," in Albert E. Utton and Daniel H. Henning, eds., *Environmental Policy: Concepts and International Implications* (New York: Praeger Publishers, 1973), 142; and Lynton Keith Caldwell, *International Environmental Policy* (Durham, NC: Duke University Press, 1984), 9. Jessica Mathews contends that the environment did not get "catapulted from the quiet netherworld of 'other' concerns to a place among international priorities" until the 1980s; see "Introduction and Overview," in Jessica Tuchman Mathews, ed., *Preserving the Global Environment: The Challenge of Shared Leadership* (New York: W. W. Norton and Co., 1991), 15–16.

66. "We would have difficulties pleading innocence or lack of experience in this field today, considering the amount of international know-how already available"; see Peter H. Sand, "International Cooperation: The Environmental Experience," in Mathews, ed., *Preserving the Global Environment*, 239.

Glossary

ACOPS   Advisory Committee on Oil Pollution of the Sea

CAAM   Centre Administratif des Affaires Maritimes of the Memorandum of Understanding on Port State Control

CLSP   Center for Law and Social Policy

COW   crude oil washing

FOEI   Friends of the Earth International

GESAMP   Joint Group of Experts on the Scientific Aspects of Marine Pollution

ICS   International Chamber of Shipping

IMCO   Intergovernmental Maritime Consultative Organization

IMO   International Maritime Organization

INTERTANKO   International Association of Independent Tanker Owners

IOPP   International Oil Pollution Prevention certificate

l/m   liters per mile

LOS   United Nations Law of the Sea Convention

LOT   load on top

MARPOL   International Convention for the Prevention of Pollution from Ships

MEPC   Marine Environment Protection Committee

MOU   Memorandum of Understanding on Port State Control

MSC   Maritime Safety Committee

mta   million metric tons per annum

NGO   nongovernmental organization

OCIMF   Oil Companies' International Marine Forum

ODMCS   oil discharge monitoring and control system

OECD   Organization for Economic Cooperation and Development

OILPOL   International Convention for the Prevention of Pollution of the Sea by Oil, 1954

OPEC Organization of Petroleum-Exporting Countries

ORB oil record book

ppm parts per million

Protocol 1978 Protocol to the International Convention for the Prevention of Pollution from Ships, 1973

PSC Port State Control

ROB retention on board

SBT segregated ballast tanks

SCMP Subcommittee on Marine Pollution

SCOP Subcommittee on Oil Pollution

tcc total cargo-carrying capacity

TSPP Tanker Safety and Pollution Prevention Conference of 1978

UNCED United Nations Conference on Environment and Development

UNDP United Nations Development Programme

UNEP United Nations Environment Programme

Index